DATE DUE

Food Bites

Food Bites

The Science of the Foods We Eat

Richard W. Hartel
AnnaKate Hartel

Copernicus Books
An Imprint of Springer Science+Business Media

Published in the United States by Copernicus Books,
an imprint of Springer Science+Business Media.

Copernicus Books
Springer Science+Business Media
233 Spring Street
New York, NY 10013
www.springer.com

Library of Congress Control Number:
2008926673

Manufactured in the United States of America.
Printed on acid-free paper.

9 8 7 6 5 4 3 2 1

ISBN: 978-0-387-75844-2   e-ISBN: 978-0-387-75845-9

Acknowledgments

We would like to thank the following people for their assistance in making sure that the details are as correct as possible.

Katie Becker

Michelle Frame

Lynn Hesson

Barb Ingham

Steve Ingham

Liz Johnson

Barb Klubertanz

Katie Kolpin

Bob Lindsay

Kirk Parkin

Gary Reineccious

Jeanne Schieffer

Ed Seguine

Tom Shellhammer

Derek Spors

We would also like to thank the editors in the Department of Life Science Communications, University of Wisconsin-Madison, for their careful reading each month. The monthly efforts of Bob Mitchell, Katie Weber, and especially Bob Cooney are greatly appreciated. Thanks also to Shiela Reaves for her words of encouragement and sharing her approach to editing.

Special thanks also go to Linda Brazill, Features Editor for *The Capital Times*, Madison, WI.

Contents

Contents

1

What Is Food Science?

If you don't finish your dinner, you'll go to bed hungry! Parents have it right, you have to eat. Even before our forefathers crawled out of the ocean, food was an important part of life. Back then, one of our daily battles was finding the food we needed to survive, at least that is when we weren't fending off bigger predators who, in turn, wanted to make a meal out of us. Like the fish in the proverbial food chain, we were looking for smaller fish to catch while the bigger ones were chasing our tails.

Finding food has always been like that, or at least until modern times.

Before civilization, or at least before TV dinners, the food chain was much different than it is now. Most of us were farmers, hunters, and gatherers. We planted wheat, corn, and other crops to harvest, and hunted for game and local produce (berries, nuts, etc.) to round out our diets. Everyone was directly involved in finding food for survival, in one way or another.

As towns formed and grew, more and more people became dependent on others for their food supply. In return for providing such needed skills as dentist and blacksmith, maybe even banker, lawyer, and used car salesman, city dwellers got the food they needed from local farmers.

As towns and cities grew into metropolises, the connection with the farmer decreased even more, to the point where most urbanites today probably would not even know a farmer if they ran one down in their sport-utility vehicle. Most of us have never even been to a real farm. The few remaining farmers, less than 1 percent of the population, must provide food for nearly the entire population, and do it without thanks. Since the majority of us have become so accustomed to having everything we need at our fingertips, or at least at the grocery store, regardless of the season.

Our food dynamics have changed considerably in the past 100 years or so. From big cities to small towns, food is mostly purchased these days at grocery stores or supermarkets, unless we're in a hurry – then we stop at a convenience store. The biggest change in our food supply is the convenience. We expect everything to be there the minute we want it.

The foods available at grocery stores have changed significantly over the past 100 years or so too. Sure, we can still buy the basic raw materials to cook our own meals – flour to bake our own bread and whole vegetables to make our own salads. But most of us do not. When was the last time you had to slaughter, gut, and clean a turkey for Thanksgiving dinner? Mostly, we get foods that have been conveniently processed to make preparation as simple as possible.

Over the years, many of the steps in food preparation have moved from our kitchens to the processing plants. From sliced bread for sandwiches to shredded lettuce for salads, the processing industry has continually evolved to make our lives easier.

We generally take the abundance of a convenient, safe food supply for granted. But, how are the raw materials converted into the foods we eat and who is responsible for the foods we find in the store? The farmers only supply the raw materials – someone else has to turn those ingredients into convenient foods. Someone has to grind the flour and bake the bread. Someone has to shred the lettuce and make sure it is safe and that it lasts at least a week in your refrigerator.

Typically, it's people trained in Food Science who are responsible for supplying the abundance of safe and nutritious foods found on the store shelves. Food Scientists are the people who make sure our food supply is safe, convenient, and long-lasting, yet still as nutritious as possible.

Food Science is an applied field, where numerous disciplines like chemistry, physics, engineering, biochemistry, microbiology, and even psychology are applied to the production and preservation of foods. In contrast to cooks and chefs, whose main interests are in the kitchen, Food Scientists are concerned with the large-scale production of high-quality nutritious foods that are safe for consumption, particularly after extended times of storage.

Like the Twinkie, which despite its dubious nutritional status has a shelf life of over two years.

Sure, you might question how a Twinkie fits into the grand scheme of food production. Actually, as opposed to our ancestors, some of whom ate bark to survive, we now eat for more than simply nutritional purposes. We also eat for psychological reasons. A cream-filled cake may be an important component of the diet for some of us. Other than the hard-core, most of us periodically eat decadent treats for psychological enjoyment rather than nutritional needs. Furthermore, in general, as long as we temper our enjoyment of these treats with healthy doses of nutritious foods, there is no harm in it.

In the following chapters, various aspects of our food production are explored. From food safety to Pop Rocks, we will delve into the science that goes into our food supply.

2

Processed Foods: Good or Bad?

How much time is spent in your home on food preparation?

Over the years, the time spent on home food preparation has decreased as our lives have become more hectic. It used to be that someone, usually mom, would spend all day, or at least a few hours, preparing the family meal. Other than special occasions, no one spends that much time on food preparation these days. Many people spend less than 15 minutes a day on meals.

One recent study[1] on food preparation times showed that one third of women and two thirds of men reported that they spent no time at all preparing food at home! Of course, some of those men were lazy husbands relying on their wives to make their meals; still, that is a large percentage of the population that does not cook at home at all.

The trend of moving some of the food preparation time from the kitchen to the food manufacturing facility started a long time ago. The first cereals, developed in the mid-1850s, reduced the time for breakfast preparation. The TV dinner of the 1950s strengthened that trend, as the availability of frozen foods skyrocketed. That trend continues these days, for example, with microwavable frozen dinners and grocery store take-out counters.

The trade-offs for the time and convenience of transferring food preparation away from the kitchen are many. Taste, for example. A well-cooked meal prepared "from scratch" generally tastes better than a pre-made, microwave-heated dinner from the freezer. So far, no

---

[1] http://www.atususers.umd.edu/papers/atusconference/posters/JabsPoster.ppt (retrieved 8/21/07)

one has come up with a replacement for mom's traditional holiday Turkey Dinner. For many of us, however, this loss in quality is worth the greater convenience.

However, prepared foods do not always mean lower nutritional value. Food processors use methods of preparation carefully designed to minimize nutritional losses while maximizing food safety. In some cases, processing methods may even lock in nutrition before the food can go bad. For example, frozen vegetable manufacturers claim that their products are frozen within a couple of hours of harvesting. The frozen peas in your freezer may have a vitamin content equivalent to that of the peas picked fresh from your garden.

Processed foods often cost more than the prepared food if we bought the ingredients and prepared the food ourselves, but not always. Large food processors get discounts for buying huge quantities and can pass this savings onto consumers. The cost of a boxed cake mix, where you just add eggs, water and oil, is probably less than what it would cost to purchase the individual ingredients. Especially since you cannot just buy the small amount of ingredients you need for one cake.

Furthermore, some foods are not edible in their native form and must be processed into something we can eat. No one nibbles on raw wheat kernels; yet, wheat is one of our food staples. After being milled into flour, wheat is turned into bread, cake, and numerous other products. Wheat is also one of the primary ingredients in cereals, which are probably one of the earliest examples of processed convenience foods.

The first commercial breakfast cereal supposedly was developed by William Kellogg at a sanitarium in Battle Creek, MI, US, as a healthy and convenient start to the day. Although the first trial batches were made in their kitchens, over the years, cereal manufacturing plants have gotten larger and larger to meet the growing demand.

Nowadays, at an average-sized cereal processing plant, about 500 tons (that is a million pounds!) of raw grains and flour are brought in each day in railroad cars and trucks, and converted into thousands of packages of breakfast cereal.

Over the years, an endless variety of cereal products have been developed, from healthy products (that often taste curiously like cardboard) to the sugary sweet cereals favored by most kids. Just add milk for an instant meal. Preparation time is negligible. And, if you are too rushed to slice fruit into your bowl, there are packaged cereals with dried fruit already added.

If you think even pouring milk in a bowl is too much of an effort, you can buy the entire bowl of cereal, including the fruit and milk, in a convenient bar. Some of us might think these new cereal bars are over-processed, but others buy them as a convenient and nutritious breakfast they can eat on the way to work.

Regardless of your perception of the food industry, processing of foods serves an important purpose in our society – providing a variety of foods with the convenience we want and the nutrition we need.

3

Vintage Wines and Chocolates*

What do wine and chocolates have in common? Sure, a nice red wine goes well with a smooth, dark chocolate, but let's dig deeper into the raw materials. The grape and the cocoa bean, from which wine and chocolate are derived, are both plant products whose characteristics vary with each harvest. The quality of both grapes and cocoa beans, and therefore wine and chocolate, depends on environmental factors like rainfall, sunshine, and temperature, which affect the chemical composition within the raw materials.

You know from experience that not every grape is equally sweet and delectable – sometimes there are sour grapes or even grapes that look delicious, but have little flavor. The same variability is also found in cocoa beans. In fact, almost all fruits and vegetables experience some degree of variability from harvest to harvest.

Vineyards use the variability in grapes, from year to year and region to region, by making vintage wines with unique character. Wine from a good vintage year can be a lot different from the same wine made in a different year. Despite that same variability in cocoa beans, however, chocolate manufacturers generally want their chocolate to taste the same no matter what.

In fact, most food processors work with variable raw materials, yet must produce a consistent product. This variability is what makes the food processing industry unique from many other processing industries. Food manufacturers somehow must accommodate differences in their raw materials to make a product that tastes, looks, and feels the same day after day.

---

* Not published as a column in *The Capital Times*

How do the large chocolate makers account for variability in the cocoa bean to produce the same product year in and year out?

Chocolate makers have chocolate tasters, who go to the source to taste the raw materials. Sounds like a great job, doesn't it. Unfortunately, chocolate tasters taste the cocoa beans and not the finished chocolate. Chocolate liquor is ground-up cocoa beans, but despite being called liquor, there's no alcohol in it to offset the bitterness. Chocolate liquor makes your mouth pucker so badly that it makes those sour candies seem sweet. Try some yourself – its often called Baker's chocolate.

These brave chocolate tasters evaluate beans from various sources and select those beans that they know from experience will give the taste they are looking for. Chocolate makers then blend beans to wipe away differences in individual batches, and produce a consistent product. To some extent, chocolate makers can also manipulate process conditions, like roasting temperatures, to make sure their chocolate tastes the same regardless of the differences in cocoa beans.

In fact, if food manufacturers knew enough about the chemistry of their products, they could adjust conditions to offset differences in raw materials and make a consistent product. For example, grape juice manufacturers, working with the same grape variability as the vintner, are capable of producing juice with a consistent taste.

Grape juice producers use an approach called standardization, where the chemical composition of the raw material is adjusted to ensure uniformity of the important factors that affect their product. They measure acidity, sugar content, and a variety of other parameters, and then blend juices from different sources with different levels of these parameters to make a product that tastes the same all year round, despite huge differences in grape quality.

Perhaps, the day is coming when we can eat a vintage chocolate with a vintage wine. In fact, varietal or single-origin chocolates, which celebrate the intricate differences of taste of a specific bean from a specific growing region, are a growing trend.

4

Preserving Strawberries, and Other Foods

There's nothing better than the taste of a freshly picked, ripe, and juicy strawberry. They're delicious and nutritious – just don't cover them with too much sugar. Unfortunately, in most of the world, you can only get fresh strawberries for a month or two every year.

For hundreds of years, our ancestors only ate strawberries in early summer – the rest of the year they only had memories. Thanks to the wonders of modern food preservation (and the trucking industry); we can now enjoy strawberry products at any time of the year. Preservation practices are the bread and butter, with strawberry jam, of the modern food industry.

Perhaps the earliest preservation technique was salting and drying of meat into jerky and pemmican, extending the "shelf life" so there was food during lean times. When the desert nomads first made milk into cheese, they were practicing another example of preserving a perishable raw material.

However, the start of the modern food preservation industry is considered to be during Napoleon's reign, in the early 1800 s. Napoleon offered a prize to whoever could develop a method of preserving foods to feed his soldiers in their march of conquest across Europe. In response, Nicolas Appert, an inventor, developed a method of preserving foods by heating them in a sealed jar to destroy microorganisms and prevent subsequent contamination.

Appert's invention was the start of the food canning industry. Canned foods, including those eaten by Napoleon's army, don't taste a lot like the original product – canning is just another way to say cooking the heck out of a food. It's hard to believe that Napoleon truly enjoyed fine French cuisine from canned foods.

Although canning is still an important preservation tool for the food industry, its use has been reduced in recent years by the development of vastly superior preservation techniques.

Nowadays, the food processing industry has a myriad of methods to preserve foods like fresh fruits and vegetables. They can be canned, frozen, and dried; even making jam is a form of preservation. What is the ideal way to preserve strawberries? Freezing is probably the best way to preserve nutrition, flavor, and texture; but still, a frozen strawberry when thawed generally does not have the same firmness of a fresh berry. We generally sacrifice some aspects of quality in all preservation techniques.

And, how the strawberry is frozen can have a huge impact on its quality when thawed.

Warm strawberries placed in the freezer at home may take several hours to freeze, depending on the size of the container. As ice crystals form and grow, they damage the cellular structure of the berry. With slow freezing, as in the home freezer, the ice crystals first form outside the cells, leading to an osmotic water imbalance between the intracellular water and extracellular water. Water flows out through the cell wall to offset this imbalance, causing dehydration and shrinkage of the cells. Ultimately, this leads to breakdown of the cell walls and loss of structure upon thawing. The result is a mushy strawberry when thawed.

In the food processing plant, strawberries can be frozen in blast freezers, solidifying the berries within minutes and reducing the possibility of osmotic water loss from the cells. Even better, the strawberries can be immersed in liquid nitrogen to freeze within seconds. The changes to the cell structure are drastically minimized compared to slow freezing, leaving strawberries with nearly fresh-like texture upon thawing.

The food industry is always looking for new and better methods to preserve foods. One "new" way to preserve strawberries and other foods, if a process that has been studied for over 40 years can be called new, is to irradiate them with high-energy gamma rays or X-rays. Irradiation uses ionizing energy to destroy microorganisms and stop respiratory reactions in a food, leaving them safe to eat, yet unable to decay or sprout. An irradiated strawberry can last for months without losing quality (see Chapter 7).

High-pressure and oscillating electric fields are new technologies that can be used to preserve certain foods without changing taste or texture. For example, exposing foods to pressures higher than those found at the bottom of the ocean preserves them without heat (the enemy of fresh fruits and vegetables) and maintains the highest quality. Guacamole preserved with high-pressure processing has recently become available in the marketplace.

Perhaps eventually, we will be able to preserve a strawberry so that it tastes just as good in December as it does in June. But, for now it's still best to do what our ancestors did – eat as many berries right off the plant as you possibly can because they won't be quite the same until next year.

5

Science Projects in Your Refrigerator

Take a long look in your refrigerator. Much of what you need for dinner tonight is likely to be in there. Refrigerators can preserve foods for a long time, but eventually most foods go bad, leading to some interesting science experiments if foods are left in the fridge for too long. The mold growing on your cheese may provide a nice color contrast, but it's going to have to be tossed.

The variety of fresh produce, dairy products, meats, and other foods we store in the refrigerator would have astounded our ancestors. But we take it for granted, unless the power goes out. Imagine what life would be without it.

It hasn't been that long that we've had the refrigerator to store our perishable foods. It's been less than 100 years since they stopped cutting ice out of Lake Wingra in Madison, WI, to use in the ice box to preserve food. Before that, only the richest among us could store food on ice to keep it from going bad. I suppose it's good that Wisconsin has cold winters; at least for half the year we could preserve foods simply by putting them outside. What did they do in Florida?

The first electric refrigerators made their appearance in American households in 1916, although the refrigeration system had been invented in 1851 – in Florida, of course. Since electricity was uncommon until the early 1900s, it is no surprise that it took over 50 years for the refrigerator to make it into American kitchens. But within a few decades of its introduction, the refrigerator became a standard item in the American kitchen.

By 1956, about 80 percent of American households had refrigerators. In contrast, only 8 percent of English households had refrigerators then. What did the British do to preserve their foods, especially those who lived in the larger cities, and why were they so much slower to embrace this wonder of technology?

Perhaps the European approach to obtaining and preparing foods can explain the difference. Europeans traditionally buy fresh foods every day, for use in that day's meal. There's little need to store food if it's going to be used right away and none is left over. Most fresh food doesn't go bad that fast.

Americans, on the other hand, rely quite heavily on the refrigerator as a means to prevent spoilage of foods. From milk to cold cuts, the refrigerator extends the shelf life of many foods well beyond what we would find if the product stayed at room temperature. The refrigerator allows us to shop for groceries once a week and keeps the food relatively fresh before preparation.

For example, milk right out of the cow goes bad within a day or two if it's not refrigerated. Even pasteurization only extends the shelf life for a few days. With pasteurization and refrigerated storage, we can extend the shelf life of milk for over two weeks. Whereas our ancestors had to salt and dry their meat to make it last, we can use the refrigerator to slow down spoilage and extend the shelf life of our cold cuts.

Refrigeration is also used to ship many food products. Those trucks with huge air conditioners on them (often called reefers) are either refrigerator or freezer trucks designed to keep the foods at low temperatures. Reefers bring produce from the fields of California to the grocery stores throughout the country with minimal loss of quality.

What will the refrigerator of the future be like? Some suggest it will have a computer that automatically inventories the contents and alerts us when a product is about to spoil – no more science projects, such as wilted lettuce, slimy meat or milk that goes plop, plop when you pour it in your cup. We could even have the computer act like mom – "close that door, you're letting all the cold escape!"

No matter what else it does, the primary function of the refrigerator will remain the same – to preserve our foods for an extended period and to make our lives safer and easier.

6

Freeze Drying – High-Quality Food Preservation

The ancient Incan people used to store their meats and vegetables high on the slopes of the Andes Mountains in Peru. The cold temperatures froze the food and the low atmospheric pressures at the high altitudes dried the food out. The Incas were practicing one of the earliest methods of preserving foods: freeze drying.

Let's look at how two different preservation techniques, freezing and drying, come together to produce extremely high-quality dried foods.

In your freezer, you'll find all types of foods, from meats to vegetables and fruits. Freezing of foods was first developed commercially by Clarence Birdseye in the 1920s when he invented a process for quick freezing that preserved much of the quality of the original food. The problem is, you need a freezer (or a very cold mountainside) to preserve frozen foods. Without a freezer, frozen foods quickly thaw and deteriorate.

Another traditional food preservation method is drying. Beef jerky, for example, is still made using essentially the same technology the Plains Indians used to preserve meats hundreds of years ago.

Drying removes water through evaporation. Liquid water molecules in the food are converted into water vapor when heat is applied, causing the food to dry. Drying often leads to shrinkage and other undesirable changes in the food as water is removed. Drying meat over a fire to make beef jerky, for example, causes the meat to turn brown, shrink, and become tougher than gristle.

Furthermore, many of the flavor compounds in foods are lost as the water evaporates.

That leads us to freeze drying, which uses both freezing and drying to preserve foods. Although freeze drying is often used to reduce the weight of food for astronauts and backpackers, the benefits of freeze drying are most visible in the high quality of the final product. Freeze-dried coffee definitely has better flavor than regular (heat) dried coffee, a difference also reflected in the price.

The Incan people placed their meats and vegetables in the cold snow and ice of the high Andes, where much of the water in the food froze into ice. The frozen food dried at the low-atmospheric pressures in the mountains by a process called sublimation. At low temperatures and pressures, water molecules in the ice crystals sublime directly into water vapor molecules – the main principle of freeze drying.

Sublimation also occurs in our home freezers, although the process occurs very slowly at normal pressures. When sublimation occurs in the food in our freezer, our food dries out. We call that freezer burn (see Chapter 19), a problem that discolors and toughens frozen foods like meats and vegetables.

However, in freeze drying, the sublimation process is carefully controlled to maintain the quality of the food. A freeze-dried food manufacturer promotes sublimation, and thus, freeze drying, by reducing the pressure almost to a complete vacuum. Near-vacuum conditions accelerate ice to vapor sublimation, drying most foods in a matter of hours.

Freeze-dried products are riddled with numerous small holes and pores, where the ice crystals used to be. Freeze-dried foods retain their texture because, unlike air-dried foods, there is no shrinkage and the pores allow easy access for water during rehydration. Plus, all the flavor molecules stay right where they belong – in the foods, so we can enjoy them. Freeze-dried foods have superior quality – think of the flavor of freeze-dried coffee.

The down-side to freeze drying is the cost, so freeze-dried foods are generally pretty expensive.

Freeze drying can be used for more than foods, too. Did you know that pets can be preserved by freeze drying? The same principles that apply to drying the berries in your breakfast cereal apply to preserving people's cats and dogs.

Although the Incan people certainly didn't understand the science of freeze drying, they were fortunate to have such an excellent food preservation method at their disposal.

7

Does Your Food Glow
in the Dark?

Despite the image of a radioactive hamburger patty lighting up the dark, irradiated foods have been proven safe in so many studies that NASA feeds them to astronauts on space missions.

In fact, food irradiation has been studied since 1905, when the first patents were issued on killing bacteria in foods with ionizing radiation. Based on numerous studies since then, the FDA has approved irradiation for use on a wide range of foods, from hamburger patties to strawberries to sprouts.

Besides killing microorganisms, irradiation also kills pests like fruit flies and insects in foods, and inhibits sprouting of vegetables. Even though irradiation has been approved for many foods, its use in foods is limited. The main application of irradiation is for medical supplies – it is widely used to sterilize items like eye drops and band-aids.

One of the arguments against irradiation of foods is that it causes changes in the food. Yes, that's true, but so does canning, and so does every other food processing method. In fact, some Food Scientists have stated that if canning had received the same level of scrutiny as irradiation, canning would never have been approved. Even something we now take for granted, pasteurization of milk, supposedly took 50 years before it was accepted as a safe processing technology.

Let's look at the sources of irradiation and then address what they do to foods.

The first irradiation source for foods was radioactive cobalt-60 (the atomic number), which emits gamma rays, a highly energetic form of electromagnetic radiation. Gamma rays, discovered first in 1900 by French physicist Paul Villard, are emitted as the radioactive cobalt-60 decays into stable nickel-60. As a radioactive material, cobalt-60 continuously emits gamma rays; to turn it off, the source has to be shielded, usually in a pool of water.

Two other sources of high-energy electromagnetic radiation – electrons and X-rays – are also used to irradiate foods. These are simple on–off systems that have the same effect as cobalt-60, but without the radioactivity issue. In fact, electron beam irradiation uses a technology similar to that used in cathode ray tubes, found in old televisions and computer screens, except with a higher energy level.

When electromagnetic radiation impacts a food, photons of energy affect food components in two ways. First, there is a direct effect that causes immediate damage, with the biggest effect on larger molecules because the photons are statistically more likely to impact these larger molecules, like DNA. One estimate quotes that a 1 kGy dose, a moderate level of irradiation, causes about 14 double-strand breaks for an *Escherichia coli* DNA molecule – enough to ensure that it does not replicate and the cell dies. It is exactly this effect on DNA that makes gamma radiation so harmful to humans.

There are also secondary effects of ionizing radiation. The photons impact molecules like proteins, fats, carbohydrates, and even water, one of the smallest of molecules. The gamma rays induce ionization of these molecules, driving off an electron. Radiolytic products, including free radicals, are formed that migrate around the food causing secondary effects like flavor degradation and nutrient destruction. These effects are quite small at the allowable radiation doses, and have little effect on the quality of the food.

Despite being approved for a range of foods, there are very few irradiated foods available, mostly because of lingering negative consumer perception. Every once in a while a supermarket stocks irradiated strawberries or mangoes, but it's quite rare. If you look hard, you may find irradiated hamburger patties, an application approved due to concerns of *E. coli* contamination.

All irradiated foods must be labeled with the Radura, which is supposedly a representation of a plant inside a package irradiated from the top (hence, the holes). Even if you haven't purchased anything with the Radura on it, you've probably eaten food that contains irradiated materials. About a quarter of the spices used in the US have been irradiated and their use in foods does not have to be declared on the label.

But don't worry, astronauts have been eating irradiated food since the Apollo missions landed on the moon. NASA sees irradiation as a tool for providing safe and nutritious foods on extended space voyages. And the astronauts say the food is delicious.

8

Is Your Food Safe?

When I was growing up, eating cookie dough from the bowl was a treat, as long as mom wasn't watching. Raw cookie dough won't give you worms, like one mom said to keep her kid's hands out of the bowl, but it certainly has potential to cause health problems.

Nowadays, we know that cookie dough with raw eggs is a likely source of *Salmonella* and a potential source of foodborne illness. Even though the potential for contamination of a raw egg is quite small, it is best to be safe and only eat the baked cookies. The high temperatures of baking are sufficient to ensure the safety of the cookies, even if the raw dough was indeed infected.

Other foods with potential contamination problems include undercooked hamburgers, which are risky thanks to *E. coli* from raw ground beef, and raw milk, which is off limits because of *Listeria*. Even raw spinach and ice cream can be contaminated, as shown in recent recalls.

Maybe, it seems that our food supply is less safe now than it was in the past, but it only seems that way because of the headlines. Any food recall gets media coverage; on the other hand, "Millions of chocolate chip cookies and ice cream cones consumed uneventfully today" doesn't grab many headlines.

In reality, our food supply is extremely safe. Of the millions of food products purchased in the grocery stores each year, only a very, very small fraction causes problems. Our food supply is undoubtedly the safest its ever been in history.

The food processing industry goes to great lengths to ensure that the foods we buy are safe. State (WI Department of Agriculture, Trade, and Consumer Protection) and federal (FDA and USDA) agencies have regulations for safe manufacture of foods and inspect all food plants on a regular basis.

Take Cookie Dough ice cream as an example. The raw ingredients coming into the plant, particularly the cookie dough bits, are tested to make sure they comply with all specifications. The cookie dough bits are blended into the partially frozen ice cream before hardening to freezer temperature. Thus, the cookie dough bits need to be made with pasteurized eggs in order to be safe. They would be tested upon arrival at the ice cream plant to ensure they contained no harmful bacteria.

The products are then processed according to Good Manufacturing Practices. For example, the raw milk is pasteurized, by regulation, so that it is held at elevated temperature for sufficient time to ensure that all harmful microorganisms are destroyed. Even more, the thermometer used to measure pasteurization temperature must be calibrated periodically to ensure that the level of pasteurization is correct or else the milk cannot be used.

Why are foods still contaminated then? One of the most likely sources of contamination is the people who work in the plant. One of the most extreme examples was Typhoid Mary, who was a New York City cook in the early 1900s. As a typhoid carrier, she infected people through the food she prepared. To prevent such a scenario from happening in their plants, the food industry works diligently to make sure that people do not cause contamination.

Thus, everybody who enters the plant must follow specific safety precautions and sanitary procedures, including washing hands thoroughly (at least 20 seconds) and wearing hair nets. The guys with beards are even required to wear beard nets – no hair, or anything else, shall get into the food. All employees remove jewelry, watches, and anything else that might fall into the food. Entry to the plant is restricted and requires shoe sanitation by, for example, walking through sanitizing foam to destroy microorganisms on shoes (dirt is a good source for microbes).

If you think getting into Fort Knox is difficult, try getting into a food manufacturing plant some time. The food industry goes to great lengths to make sure the people who work in the plant cannot contaminate the foods.

Even the design of the manufacturing plant and equipment are done with sanitation in mind. Raw material areas are kept separate from product areas to prevent post-processing contamination, and all equipment are designed to be easily cleaned and sanitized.

To ensure that all products leaving the plant are not contaminated, a statistical sampling of product is tested every day. A dedicated quality control manager ensures that all products, including Cookie Dough ice cream, are always safe to eat and enjoy.

Although these extensive precautions generally prevent contamination of our food supply, accidents may still occur. Human error or oversight is often the root cause of most food contaminations.

Rest assured that the vast majority of the foods you purchase were made according to Good Manufacturing Practices and pose no risk for foodborne illness as long as they are handled and prepared properly. You can enjoy that Cookie Dough ice cream cone knowing it's safe to eat.

9

Food Safety and Mobile Food Carts

I'll have a chicken stir-fry, thank you, but hold the *Salmonella*. Although it may taste delicious, how do you know your lunch at that downtown cart is safe to eat?

Thanks to our state food inspectors, we should neither worry about eating lunch at the carts downtown, nor should we worry about the food at the grocery store deli or the cafeteria at work.

In Wisconsin, the state Department of Health and Family Services, Division of Public Health, is responsible for developing regulations and controlling how the food is served at restaurants, cafeterias, and even downtown food carts. Each establishment that sells food to the public must be licensed by the state and is subject to periodic inspections. The regulations clearly define the hygienic practices that must be used when serving food to the public.

In fact, nearly all food facilities, whether cafeterias and restaurants or seasonal carts (mobile food establishments), must employ a person with "Wisconsin certified food handler certification." This person ensures that all state regulations are met and the food is safe to eat.

To become a certified food handler, a person must pass a test that demonstrates their knowledge of safe food handling practices. Besides knowing how to wash hands properly, a certified food handler must know how to properly store raw materials, handle and prepare foods safely, and hold foods ready to be served for hours without incubating microorganisms. We expect the lunch from a downtown cart to be as safe as any packaged food we buy at the grocery store. It is the responsibility of the certified food handler at each establishment to see that it is.

The Division of Public Health says one common mistake at mobile food service establishments, perhaps the mistake most likely to cause illness, is the temperature at which the food is held prior to serving. Foods must either be held at high enough temperatures (above 140°F) that microorganisms cannot grow or be sufficiently cold (below 40°F) to inhibit microbial growth.

Intermediate temperatures, around room temperature or slightly above, provide an ideal climate for bacteria to grow, and must be avoided. Inspectors check to ensure that food holding temperatures at all food vending sites are within the acceptable range. If not, the establishment may be closed down until the problem is rectified.

Are your food-handling practices at home up to the same standards as the downtown carts? We should all be certified food handlers, but a recent study suggests that we are not always as careful at home as we should be. Do you put left-overs directly into the refrigerator at the end of the meal or do they sit out on the counter for some time?

We've learned a lot about kitchen safety over the years. My mom used to let the uneaten roast and gravy sit out on the kitchen table until they had cooled to room temperature (it is easier on the refrigerator). Now, we know that microorganisms love to grow in food left on the counter for too long. Leave your food out too long and you may get food poisoning. It's a wonder that we didn't get food poisoning more often – or did we, since it's often experienced as nausea and diarrhea.

How about your cutting boards? Do you use different boards for handling raw meats and raw vegetables, or at least wash the board well (with soap) between handling these products? Cross-contamination of vegetables by raw meats is a common food safety mistake at home.

These problems, and more, are the things that a certified food handler knows to avoid and the state inspectors check during their annual (and sometimes more often) visit to each food establishment.

For the most part, thanks to our state health inspectors, your lunch from a downtown cart is not only delicious, it's also safe to eat.

10

At Work in a Vale of Tears

"Life is like an onion. You peel it off one layer
at a time; and sometimes you weep."

Carl Sandburg,
American poet

You know what it is like to be in the crying zone when you cut a raw onion in the kitchen. Now imagine cutting 70 tons of onions – a typical day's work at a local snack food manufacturing plant, where truckloads of raw onions are turned into frozen, deep-fried onion rings. That's a lot of onions, and a lot of tears. At the plant, they say you get used to cutting onions and stop crying after a few minutes.

As difficult as it may seem to get used to crying every day, this is the domain of the quality control (QC) manager of the plant. He/she oversees day-to-day operations to make sure the breaded onion rings come out the same each day. And now, he/she only cries when something goes wrong in the plant.

The QC manager's job at any food manufacturing facility is twofold: first and foremost, to ensure that the products are safe to eat, and second, to ensure that they have the highest quality. At the frozen onion ring factory, onions brought into the plant from delivery trucks are properly cleaned, sorted, and peeled before being carefully sliced into rings – this is the real crying zone. The onion slices are dunked in the batter before being sent through a deep fat fryer. Finally, the partially cooked onion rings are frozen prior to packaging.

As a Food Scientist, the QC manager applies the principles of chemistry, physics, engineering, microbiology, business, and even psychology (why do you eat what you eat?) to the manufacture of foods on a large scale. In contrast to cooks, whose main interests are in the kitchen, Food Scientists are concerned with the large-scale production of high-quality, nutritious foods that are safe for consumption, especially over extended periods of storage.

A home cook might clean, sort, peel, and cut a pound of onions to make onion rings in the kitchen; a chef might process 20 pounds a night in a restaurant. The manufacturing plant clears 140,000 pounds of onions every day. That gives you an idea of the scale of the job.

The QC manager's most important job is to ensure that the finished product is essentially free of bacteria, so that it is absolutely safe to eat. But more than that, the QC manager must ensure that the product has the highest quality standards. In the case of frozen onion rings, quality can mean a lot of things. They need to have a full coating of batter – no half-dipped rings allowed. Each box must be filled with exactly the right amount of onion rings – too little and the consumer (not to mention FDA) complains, yet over-filling means money lost. These concerns, and many more, are part of the QC manager's daily concerns.

What causes you to cry when you cut open an onion? Cutting an onion breaks open the cells, allowing an enzyme to react with amino acids in the onion. Various sulfur compounds are produced, including a volatile compound called *syn*-propanethial-*S*-oxide. This volatile compound vaporizes into the air and gets into your eyes, where it forms sulfuric acid. You'd cry too with sulfuric acid in your eye.

However, at the onion factory, they build up immunity to crying over cut-up onions. Somehow their eyes adapt to the acid environment and stop tearing. You can quickly tell in the plant who the new people are – they're the ones with wet eyes.

Short of working at the onion factory to build up immunity, how can you cut onions without crying? Numerous techniques have been suggested, from cutting the onions under water to wearing goggles.

Some even suggest chewing on a slice of bread as you cut the onions so that the vapors are absorbed into the bread instead of your eyes.

One cook recommends carefully cutting out the bulb of the onion at the root side since the bulb is where the enzyme is concentrated. If you don't cut open the bulb, you won't cry when cutting the onion. Or you can wait for onions to be genetically engineered to remove the problematic enzyme – the tear jerker onion.

Until then, the QC person at the onion plant must maintain tear immunity to sliced onions while making sure his/her product meets all sanitary and quality standards.

11

Are All Microorganisms in Food Bad?

What gives Limburger cheese its delightful aroma? It's the choice of bacteria.

Bacteria in food? How can that be? "Contains active cultures" – that's what the label on the yogurt container reads. And, "starter culture" in summer sausage is a blend of bacteria that helps to preserve and add a tangy flavor. Yogurt, sausage, and Limburger cheese are examples of foods that are made with the benefit of microorganisms.

Despite growing concerns over foodborne illness and microbial contamination of our food, not all microorganisms in our foods are bad. Numerous microorganisms, like bacteria, yeasts, and even molds, are used in a wide variety of foods. In fact, they provide some of the most important and even enjoyable aspects of those products.

Without microorganisms, we wouldn't have cheese and yogurt, bread, pickles and sauerkraut, and certain sausages, not to mention beer and wine. Without microorganisms, Swiss cheese wouldn't be holey and Bleu cheese wouldn't be moldy. And Limburger cheese wouldn't be smelly.

The microorganisms consume certain nutrients found in the food, with their products giving rise to specific effects. For example, yeasts consume the sugars in grape juice to produce alcohol in wine. Yeasts also consume the sugars in wort, the liquid obtained from extracting malted barley, to produce alcohol in beer. In fact, it's the different types of yeasts that result in the variety of beers that are available. For example, *Saccharomyces cerevisiae*, a top fermenting yeast, is used to make ale, whereas *Saccharomyces carlsbergensis*, a bottom fermenting yeast, is used to make a lager.

Many of these microorganisms also produce carbon dioxide, necessary for the bubbly character of champagne and beer.

Perhaps, one of the first fermented products was sour cabbage, or sauerkraut. The mixed microbial flora normally found on the cabbage is all that's needed for fermentation; you can tell that by the smell of an unharvested field of cabbage. Whew! Commercial sauerkraut production uses a little salt to restrict growth of certain microorganisms and control the fermentation. Preservation of cabbage as sauerkraut even has numerous health benefits – it is thought to be the reason why Dutch sailors didn't get scurvy on long explorations. More recently, components of sauerkraut (and cabbage) have been shown to inhibit certain types of cancer.

Many dairy products are fermented. Bacterial cultures utilize lactose in milk to make lactic acid, which causes casein to coagulate to make yogurt. Furthermore, some populations of microorganisms in yogurt actually promote digestive health, and a whole segment of the food industry is seeking to incorporate these health-promoting microorganisms, or probiotics, into our foods (see Chapter 12).

In cheese, certain types of bacterial cultures are added to help promote "ripening" during storage. During aging of cheese, a complex series of reactions take place as the bacteria degrade components like protein, fats, and sugars to generate the desirable flavors of ripened cheese. An aged Cheddar cheese develops its flavor and texture because of the ongoing activity of the microorganisms during storage. Other bacteria produce gas to make the holes in Swiss cheese. Certain molds are added to enhance flavor development in cheeses like Brie, Camembert, and Bleu.

What gives Limburger cheese its pungent aroma? It's all in a bacterial culture used during its production, in this case *Brevibacterium linens*. The rind of the cheese is washed with bacterial culture, which breaks down the proteins in the cheese to produce sulfur-containing chemicals that bear the distinctive odor characteristic of Limburger. Because its a rind-washed cheese, much of the distinctive odor is on the surface, with the interior containing far less of the smell.

Interestingly, a close cousin of the bacteria used for making Limburger cheese, *Brevibacterium epidermidis*, grows between the toes. It should come as no surprise then that Limburger cheese reeks a little like smelly feet.

Limburger cheese notwithstanding, not all microorganisms in food are bad. Enjoy a lunch of yogurt, bread, and cheese, washed down with a glass of beer or wine for one of the best microbially based meals. Such a meal may also have considerable health benefits, although as always, everything in moderation, especially the Limburger cheese.

12

Probiotics – The Growth of Cultured Foods

When the Spanish conquistadors invaded the Aztecs in the early 1500 s, they should have brought yogurt with them.

During their trip to the New World, many conquistadors developed traveler's diarrhea, sometimes called "Montezuma's revenge" after the Aztec leader who was defeated by Cortez. Their travel problems might have been reduced, if not completely prevented, if they'd had yogurt's good bacteria fighting the water's bad bacteria.

It's widely accepted that eating yogurt is good for you. From promoting good digestion to boosting immune response, yogurt is considered by many to be a health food.

At least in part, this healthy image is related to the bacteria present in yogurt. The yogurt container's label must read "Contains active cultures" if some of the health benefits, particularly for promotion of good digestion, are to be conferred.

Bacterial cultures, such as *Lactobacillus bulgaricus* and *Streptococcus thermophilus*, are added to milk to produce a protein gel that gives yogurt its semi-solid characteristic; the bacteria also contribute to yogurt's characteristic flavor. The bacteria utilize the lactose in milk to grow and reproduce, producing lactic acid in the process. The acidic conditions (reduced pH) from lactic acid production cause the milk protein, casein, to coagulate.

The result is yogurt's characteristic gel-like texture. By control of bacterial culture and processing conditions (temperature, stirring, etc.), anything from firm cup-set yogurt to soft, pourable yogurt can be produced. Kefir, a fermented milk product that is gaining in popularity, has a fluid yogurt-like consistency due to the type of bacteria and yeast used in its production.

Besides helping to make yogurt from milk, the bacteria are thought to provide health benefits, including improved digestive health and immune response. Some companies even put in more bacteria, like *Lactobacillus acidophilus* and *Bifidobacterium* species, after yogurt fermentation to enhance these health benefits. Microbial cultures that impart health benefits are called probiotics.

The term probiotics comes from the Latin roots *pro* and *bios*, which together means "promoting life". According to the National Yogurt Association, probiotics are "living organisms, which upon ingestion in sufficient numbers, exert health benefits beyond basic nutrition." To give a health benefit, yogurt needs to have active cultures, something not all yogurts contain.

In fact, the National Yogurt Association requires that yogurt must have 100 million viable bacteria per gram at the time of manufacture in order to carry their "Live Active Culture" seal. Yogurts with less than this number of bacteria don't confer sufficient health benefits and so don't warrant their seal of approval.

Probiotics are thought to work by influencing the natural microbial population in the digestive system. The ingestion of live bacteria can influence the growth of the native bacteria in the intestinal tract and thereby confer health benefits. Probiotic bacteria are particularly useful when some of the beneficial bacteria in the digestive tract have been knocked out, as may happen when antibiotics are used to treat infections (of bad bacteria). Probiotics may also be useful when there are too many bad bacteria in your system, as is the case with traveler's diarrhea.

It is not just yogurt companies that are cashing in on the probiotic trend. In fact, a whole segment of the food industry is seeking to incorporate these health-promoting microorganisms into our foods. From cereal containing *Lactobacillus* cultures (and broccoli extract!) to chocolate wellness bars "containing over five times the live active cultures of yogurt", probiotic foods are being marketed with the expectation that they will help improve health (and sell well).

Unlike the Spanish conquistadors, modern travelers have numerous options for fighting traveler's diarrhea. On your next trip abroad, consider eating yogurt with active cultures or any of the new probiotic products to help head off digestive problems.

13

How to Keep Guacamole from Turning Brown

How do you stop guacamole from turning brown? Eat it all right away, of course. But what if your guacamole eyes are bigger than your stomach, or you want to prepare it ahead of time for a party. How do you keep it from turning browner than river mud?

Let's start by looking at what causes browning to occur.

Some produce, such as avocadoes, potatoes, and apples, turn brown rapidly after they've been cut open. They're fine until the minute you slice them open, then they quickly turn an unsightly color.

Slicing apples or mashing avocadoes exposes the insides of the cells to air. This allows the enzyme, polyphenol oxidase (PPO), contained within the cells to react with oxygen in the air. This enzymatic reaction leads to the formation of melanoidin pigments. With guacamole, the result is a very unappetizing, brownish-green mess.

In some cases, enzymatic browning is desired. Raisins have a deep brown color in part from PPO activity. As the grape dries, some of the cells are broken open, exposing the PPO to the air. Dark raisins are the result.

But with guacamole, this reaction is definitely a bad thing. Numerous methods have been proposed for preventing guacamole from turning brown. Some swear by putting avocado pits into the guacamole, while others say wrap it tightly. Liberal use of lemon juice is supposed to slow down browning. One chef even recommends that certain oils prevent browning of guacamole for up to three days.

But, which method is best? To find out, let's look at each method and see how well they work.

Perhaps the oldest and most commonly suggested remedy for brown guacamole is to put the intact pit in the middle of the finished product. For this to work, the pit would have to somehow interact with the PPO to inhibit browning. I tried this at home, putting the pit into the center of a small bowl of guacamole and leaving it in the refrigerator over night. The next day I had brown guacamole. Only the part directly under the pit was still green. The guacamole looked exactly the same as the control, a similar bowl of guacamole without the pit. Apparently, the pit trick doesn't work, unless you have enough pits to cover the entire bowl.

Next, I covered and sealed a bowl of fresh green guacamole with plastic wrap, leaving no air space between plastic and guacamole. I also filled a plastic bowl with guacamole and sealed it with a lid tightly, making sure there were no air spaces between guacamole and the lid. Both of these bowls of guacamole were still green the next day. Why? Because the well-sealed containers prevented the PPO from being exposed to air; hence, no reaction.

What about lemon juice? All the guacamole in these tests had a small amount of lemon juice added, so by itself the acid in the lemon juice was unable to stop the enzymatic reaction when the guacamole was exposed to air. I tried sprinkling lemon juice liberally on the top of another bowl of guacamole. This sample also had turned brown by the next day, but maybe a little less than the control. Ascorbic and citric acid in lemon juice are known to inhibit enzymatic browning, although pure ascorbic acid is supposed to be the best.

Recently, packaged guacamole in sealed pouches has become available in the market. Vacuum packaging extends shelf life by excluding air, but some brands go even further. In these brands, the PPO has been inactivated by high pressure, a process that causes the enzyme to unfold, thereby eliminating its activity. Even when left overnight exposed to air in an open bowl in the refrigerator, the high-pressure treated guacamole stayed as green as it was when it was made. The combination of inactivating PPO with high pressure and reducing oxygen content by vacuum packaging makes it possible to store guacamole for several weeks without unsightly browning.

Next time you need to store guacamole for hours or even days, try some of these more successful tricks. Through knowledge of Food Science, you can prepare your guacamole today and enjoy it tomorrow.

14

Churning the Butter*

The words of his mouth were smoother than butter, but war was in his heart.
    Psalm lv. *21*.

What is it about butter? Another slice of bread ruined trying to smear butter right out of the fridge. It's impossible to spread. Why can't they make a butter that's spreadable when right out of the refrigerator, like margarine?

Well, they do. Spreadable butters are available, sort of. But, before talking about how to make butter spreadable (the next chapter), let's talk about where butter comes from and how it's made.

Butter is almost as old as history, if anything can be that old. We know butter comes from cows, of course, but it can also come from yaks, goats, sheep or camels. In fact, it can come from any mammal's milk that contains sufficient fat. Cow's milk contains only 3–4 percent fat – it makes a nice butter. Yak's milk contains 5–7 percent fat – the high fat content makes an excellent butter. Seal milk contains over 50 percent fat – seems like it would make a great butter, but have you ever heard of seal butter? Baby seals must need a lot of fat for both nutrition and protection from the elements.

In fact, milk, including the fat, is specially designed by the animal to provide the nutritional needs of the young. Because its melting point is below body temperature, milk fat right out of the mammary gland is melted, in liquid form. This allows the milk to flow easily and provides an easily digested nutritional source for the young. Also, the milk fat provides the range of fatty acids needed by the young to grow and develop.

---

* Not published as a column in *The Capital Times*

Unfortunately, milk fat wasn't necessarily designed to spread on bread directly out of the refrigerator since it solidifies substantially when cooled to refrigerator temperatures. Milk fat, with a melting temperature of about 95°F, just below body temperature, crystallizes when it is cooled. The further it's cooled, the more fat crystallizes and the harder it gets.

But, how butter is made can influence how the fat crystallizes and therefore, affect the hardness and spreadability.

To make butter, the fat is first creamed off the milk and then churned. Cream is an oil-in-water emulsion containing about 55–60 percent water. Numerous small droplets of milk fat are dispersed in the aqueous phase. Butter, which contains about 18 percent water, is a water-in-oil emulsion, where water droplets are dispersed in a continuous phase of fat. To make butter, it requires that the original cream emulsion be inverted (oil-in-water to water-in-oil) and the water content reduced substantially. That's done in the churning process.

Because cow's milk contains only 3–4 percent fat and butter is at least 80 percent fat, it takes about 21 pounds of milk to make a pound of butter. The by-product is buttermilk. Hardly anyone drinks buttermilk any more, so it's usually dried and used as an ingredient in various foods.

The oldest known process for making butter, practiced in the Middle East, involved pouring cream into a goatskin, hanging the goatskin from a tent pole, and swinging it around until butter was formed. The agitation caused the emulsion to invert and the buttermilk was poured off, leaving butter. The old ways were pretty energy intensive.

You can make butter at home by filling a jar with cream and continually agitating the jar. Of course, you would get tired pretty quickly of shaking the jar, maybe giving up before you inverted cream into butter. That's why, over the years, numerous mechanical churns were developed to simplify the agitation process.

Some churns used animals to rock or turn the container of cream to make butter. Others used wooden barrels in a rotating contraption that were cranked manually. Probably the best known churn, though, was the dash churn. Visualize a farm wife sitting over a tall, narrow wooden tub filled with cream, working the vertical wooden plunger, or dash, up and down. She's churning cream into butter.

Modern commercial butter churns are large stainless steel machines that churn the butter continuously, exuding a stream of semi-solid butter. Each of the steps in butter-making – cream inversion, buttermilk separation, washing and kneading, and salting – is done in sequential elements within the continuous process. Over a ton of butter can be produced per hour in these modern continuous churns.

Whether you're using a goat skin or the latest commercial churn, butter comes out much the same. Regardless of how it's made, when cooled in the refrigerator, it's too hard to spread. In the next chapter, we'll explore the factors that affect butter's spreadability.

15

What Side Is Your Bread Buttered On?*

In the previous chapter, we explored how butter was made. Now we're ready to take on spreadability. Why does butter right out of the refrigerator tear apart a normal piece of bread, but margarine doesn't?

It's all in the fat crystals – how much there are, what type they are, and how they form together. It's also important to realize that milk fat, because of its diverse chemical composition, crystallizes over a wide range of temperatures. It's melted at about 95°F, but has increasingly more crystalline, or solid, fat as temperature goes down. At refrigerator temperatures, about 50 percent of milk fat is crystalline – the rest is liquid.

Butter is made by inverting a cream emulsion into a water-in-oil emulsion, but since the cream is cold when it is churned into butter, some of the fat is already crystallized. So the first method of influencing hardness of butter involves controlling the fat crystals in cream prior to churning.

Tempering cream by selecting the appropriate temperature conditions controls milk fat crystal formation, which ultimately affects spreadability. Rapid cooling of hot cream exiting the pasteurizer promotes formation of numerous small milk fat crystals, whereas slow cooling produces fewer and larger crystals. Since the number and the size of crystals influence spreadability, it's important to get the right mix.

To enhance spreadability, cream is first cooled rapidly to about 45°F and held for a couple of hours. This promotes the formation of

---

* Not published as a column in *The Capital Times*

numerous small milk fat crystals. The cream is then warmed to about 70°F and held for a couple more hours to melt some of the milk fat crystals, leaving only the ones with the highest melting point. Churning this cream gives a relatively spreadable butter.

But cream tempering can only do so much. There's still way too much crystallized fat in butter at refrigerator temperatures, somewhere perhaps as high as 50 percent of the milk fat is solidified at that point. That's just too hard to spread. Margarine, on the other hand, is designed to have only about 15–20 percent of crystallized fat at refrigerator temperatures, depending on the type of margarine. Stick margarines may have slightly more solid fat and tub margarines slightly less.

Since the solid fat content of milk fat is too high in the fridge, other methods of making butter spreadable are needed. One approach, supposedly used in refrigerators in New Zealand for a while, is to simply keep the butter warmer. You cannot keep it out at room temperature very long before it oxidizes and goes rancid, but you can keep a butter drawer in the fridge at slightly warmer temperatures. This method works, but you need a special fridge.

The New Zealanders also came up with a way to modify the nature of the fat in butter to make it more spreadable. Because milk fat has such a wide melting range, they were able to separate the highest melting components of the fat from the lowest melting fats in a process called fractionation. The two components were then blended back together in a certain ratio to give a butter that had lower solid fat at refrigerator temperature and was therefore more spreadable. The product was commercialized in the UK, in the 1990s, but has since fallen off the market for economic reasons.

More recently, spreadable butter is available where the solid fat at fridge temperatures has been reduced with canola oil. The liquid canola oil simply dilutes the milk fat, decreasing the amount of solid fat and softening the butter.

Does this method work? Somewhat, but not that well. Perhaps, it is a little easier to spread right out of the fridge than real butter, but it is still not like margarine. Why not just add more canola oil then to make it spread easier?

The problem is at the other end of the temperature spectrum. If you take too much of solid fat away, then it won't be hard enough at room temperature. It will just flow as a liquid. It won't be spreadable, it'll be pourable. Not good. Unfortunately, butter is either too hard at refrigerator temperature or too soft at room temperature.

So, spreadable butter is still somewhat difficult to get right. Your best bet is to remember to take the butter out of the fridge 15–20 minutes before you plan to spread it on your bread. The warmer temperature reduces the amount of solid fat, making it softer and easier to spread. The problem is, who can ever remember?

16

Butter or Margarine?

Nutritional information can be quite confusing, especially when it comes to fats. Yet, each new nutritional discovery leads to a wave of new and, we hope, healthier foods as manufacturers respond to consumer concerns.

Years ago, we were told that it was better to eat margarine than butter because the saturated fats found in butter could lead to coronary heart disease. Tropical fats were also bad because they had high levels of particularly unhealthy saturated fats. In response, companies replaced tropical fats in foods and we ate less butter.

But now we're told that butter may not be so bad, especially when compared with margarine made using hydrogenated vegetable oils.

Recent studies have shown that partially hydrogenated vegetable oils, which lead to the formation of *trans* fatty acids, may be even worse for us than the saturated fats that we're supposed to avoid.

Due to these health concerns, all foods are required to have *trans* fat content, along with saturated fat content, listed on the nutritional label. Many people, and even entire cities, are taking these warnings quite seriously by completely eliminating them from their foods. Let's look at *trans* fats, what they are, and why they've found their way into our foods.

Fatty acids may be saturated or unsaturated. Saturated fatty acids are long chains of carbon arranged in a straight, almost linear fashion. Unsaturated fatty acids in foods come in two forms, *trans* and *cis*, depending on the orientation of certain carbon molecules in the fatty acid. The carbon chain of *cis* fatty acids has a sharp bend, looking roughly like a "V," whereas *trans* fatty acids contain a straight carbon chain with a slight kink in it. *Trans* unsaturated fatty acids are more like saturated fats in shape than like *cis* unsaturated fats. This difference in conformation of the two types of unsaturated fats, one kinked and the other bent, gives them different physical properties and health impacts.

For example, they have different melting points. *Cis* unsaturated fats have very low melting point, below room temperature, because it's difficult for the bent-chain molecules to come together. Thus, they're typically liquid oils. The *trans* unsaturated form, however, with a straighter chain, forms crystals more easily and has a melting point slightly above the body temperature. This higher melting point is what makes hydrogenated vegetable oils containing *trans* unsaturated fats particularly useful in shortenings and margarines. However, there's something about that kinked chain conformation that leads to the negative health impacts of these *trans* unsaturated fats.

So what fats can we eat? Unsaturated oils, like canola and olive oil, are still reasonably good for us, in moderation. They do not make very good frying oils, however, and they cannot be made into margarine or shortening without somehow modifying their melting properties.

That leads us to methods for fat modification. If hydrogenation is no longer acceptable because of the *trans* fat issue, what options do we have to make fats with the right melting properties for breads and cakes?

One recent approach is to genetically engineer plants to produce fats with more desirable properties. For example, soybeans are being genetically modified to contain fewer polyunsaturated and saturated fatty acids, leaving mostly monounsaturated fatty acids (*cis* oleic acid). This oil would not become rancid so fast, a common problem with polyunsaturated oils, and they can be used as frying oil without partial hydrogenation.

Making low *trans* shortenings and spreads is more difficult. One way to do it is to fractionate a fat like palm oil into a hard fat and liquid oil. Fractionation involves slowly cooling the melted fat until the first crystals form, and then separating off these high-melting point crystals. The high-melting fat (called stearin) contains primarily long-chain saturated fatty acids (palmitic and stearic acids), whereas the liquid oil (the olein) contains more unsaturated fats. By carefully blending the high-melting stearin with a liquid oil in the proper proportion, a spread with desirable properties (hardness, spreadability, etc.) can be obtained.

Unfortunately, this approach to reduce *trans* fatty acid content leads to an increase in saturated fats, depending on how much of the hard fat must be added. In one sense, we're stuck between the rock of saturated fats and the hard place of *trans* fats. However, if done well, the *trans* fatty acid content can be reduced to zero with only a small increase in saturated fat content. Besides, at this point, there are very few options to replace *trans* unsaturated fatty acids in our foods and still get the desirable texture and mouthfeel that we want.

The new wave of low-*trans* products in the marketplace is a direct result of an increased understanding of the health impacts of our foods. As our understanding of health and nutrition continues to grow, new foods will be developed to better meet our nutritional needs.

17

Chocolate Flavor

A friend of mine once said that if you added a little of each of the known chemical compounds together, the result would look, smell and maybe even taste a little like chocolate. His point was that chocolate flavor comes from a wide range of different chemical compounds mixed together to a perfect natural balance.

Why do different chocolates have such different flavors? For that, we need to look back into the origin of chocolate, to the cacao plantations in equatorial countries, and into the processes used to manufacture chocolate from the cocoa bean.

Cocoa pods, the fruits that grow on the cacao tree, each contain 30–40 cocoa beans surrounded by white mucilage within a fleshy exterior. The cocoa beans are the seeds from which new cacao trees grow, and, when cleaned and dried, they are also the starting point for chocolate.

Chocolate flavor originates from the chemicals initially found in the bean, in the same way as wine grapes and coffee beans contain the origins of wine and coffee flavors, respectively. There are several different types of cacao tree, each of which produces cocoa beans with different inherent flavors. The most common type of cacao plant is the Forestero, preferred for its disease resistance and robust chocolate flavor. The Criollo plant gives a milder, nutty chocolate flavor, whereas the hybrid, Trinitario, offers a complex mix of fruity and floral flavors as well as rich chocolate notes.

Environmental factors also influence the chemical make-up of the cocoa beans. Similar plants grown in different conditions – exposure to sun, rain, temperature, and humidity during the growing season – lead to chocolates with different flavors. Thus, chocolates made from Ivory Coast cocoa beans taste different from chocolates made from Costa Rican beans. Chocolate companies carefully blend beans from different sources to get the same flavor every time.

Chocolate flavor also depends on how the cocoa beans were processed. Fermentation of the raw beans and roasting of the dried beans are the most critical processes of flavor generation.

After the cocoa beans are removed from the pod, they are allowed to ferment in their natural mucilage. The fermentation generates important flavor precursor chemicals (amino acids, sugars, etc.) critical to chocolate flavor. Although fermented beans don't have much chocolate flavor themselves, these precursors are important in developing a satisfactory chocolate flavor in the finished product.

Roasting converts the precursor chemicals produced in the fermentation into desirable chocolate flavors. The high temperatures of roasting trigger numerous chemical reactions, some of which are similar in many respects to those of toasting bread. In both toast and cocoa beans, proteins and sugars react to form color and flavor compounds. It's just that the starting materials in bread are different from those in cocoa beans, so the flavors produced are different.

Conditions during roasting provide another variable. Just as coffee beans can be light roasted or dark roasted, changing the heat treatment during roasting gives different chocolate flavors. Each chocolate manufacturer selects the roasting conditions that best suit the flavor profile they want in their chocolate.

The complexity and variation of the chemical compounds in the roasted beans is what creates the full richness of natural chocolate flavor, with a complexity that is nearly impossible to duplicate. Developing a synthetic chocolate flavor that approaches that of real chocolate remains a holy grail of the flavor industry.

Flavor chemists synthesize imitation flavors by analyzing all the different component chemicals that go into a flavor, then mixing together the primary chemical components in those flavors. Many artificial flavors, such as strawberry and cherry, can be quite close to the natural version because they contain just a few main flavor chemicals.

Because of the broad spectrum of chemical components in chocolate, some of which are found at extremely low, but significant, levels, a decent imitation chocolate remains elusive. For now at least, when it comes to chocolate flavor, Mother Nature still has the edge.

18

Rice in Your Salt Shaker?

Whew, it's hot and humid today. How humid is it? It's so humid that your shirt is sticking to your body and you're just sitting still – you wish you were on a vacation at the beach. Even the salt in the shaker is on vacation, it's so sticky. Nothing comes out of the salt shaker no matter how hard you shake, and shaking so hard only makes you sweat more.

The problem is that humid air contains more moisture than many dried foods, so products like crackers, cereals, and cookies, are extremely prone to picking up moisture from the air on humid summer days. Before long, you have a mushy cookie with no snap, corn flakes with no crisp, and crackers that only a mother could love. And salt that clumps in its shaker.

Actually, water content by itself is not always a good indicator of whether or not a food gains or loses moisture. Food Scientists use a term called water activity, which is related to the escaping tendency (chemical potential or fugacity for those of you with a technical background) of the water molecules. If the escaping tendency is low, water activity is low; when exposed to humid air, a food with low water activity picks up moisture quite readily. Fortunately, water content and water activity usually go hand in hand; so when one is low, so is the other.

On the other hand, high-moisture products, like fruits and vegetables, are generally not prone to picking up moisture from the air; in fact, they suffer from the reverse problem in the winter, they dry out due to moisture loss (called desorption).

So how do grains of rice in a salt shaker prevent clumping? This has to do with the different ways that starch (in the rice) and salt crystals handle high humidity.

When the humidity is high, water molecules in the air attach to the surface of salt crystals. Above a certain relative humidity, at what is called the deliquescent point, there are enough water molecules at the surface of the salt crystal to actually dissolve the salt, forming a layer of saturated salt solution. When the humidity goes down later on, some of the water in that solution layer goes back into the air, allowing the salt to recrystallize. Since it's in contact with the neighboring salt crystals in the shaker, a crystal bridge forms between the two crystals. When enough of these bridges form, the salt clumps.

Rice starch, even though it's also dry and has a low water activity, can accept fairly large levels of water without significantly changing its properties. Any water in the vicinity is absorbed into the starchy matrix, with just a little swelling. The rice may get a little soggier, but when it dries out again, the rice generally goes back to its original state. So, it doesn't change a lot during moisture sorption. That's a useful property for the salt shaker.

A few grains of rice in a salt shaker keep the humidity in the air from dissolving the surface of the salt crystals. The rice traps water vapor molecules, protecting the salt from clumping. Except perhaps on the steamiest of days, the salt still flows freely onto our foods.

19

Frost on Your Berries

While cleaning out the freezer, have you ever found some really old food that no longer looks edible? Perhaps, some raspberries have frosted over inside the bag to form a huge berry-filled ice cube or a steak has developed an ashy, dull brown color. These foods are victims of freezer burn, the nemesis of foods stored too long in the freezer.

Freezer burn is nothing more than the food losing water. Water migrates out of the frozen food, either into the package headspace (between the food and the package) or directly through the package itself, and the results are not appealing. Freezer-burned raspberries are dry and mushy when thawed, and the brown discoloration indicates that freezer-burned meat will be tough and chewy.

But if the food is frozen, how can there be water to migrate out? In fact, even though raspberries or steaks seem to be frozen solid, there is still unfrozen water in these frozen foods. The amount of unfrozen water depends on the components found in the food and the temperature of the freezer. Foods with high sugar and salt content have more unfrozen water since these components interact with water to reduce the freezing point.

Even when a raspberry is frozen solid, down to temperatures of $-40°F$, about 20 percent of the initial water in the berry remains unfrozen. Over time and under the right conditions, this water can migrate out of the berry and into the headspace within the package to cause those berry-filled ice cubes.

Freezer burn is enhanced by normal temperature cycles in freezers. The refrigeration system that keeps your freezer cold turns on and off, much like an air conditioner on a hot summer day. When the refrigeration system is on, the freezer temperature decreases until it reaches the set point, after which the refrigeration turns off. The temperature slowly rises as heat leaks into the freezer, until at some point, the refrigeration system turns on again and the cycle starts anew.

Depending on the thermostat in the freezer, these temperature fluctuations can be as large as 5–6°F. These temperature cycles play an important role in freezer burn.

When freezer temperature goes up, the warmer air in the space between the berries draws unfrozen water out of the fruit. When temperature cycles back down again, the air in the package no longer can hold as much water and some of it condenses as frost.

Water vapor condensing on the raspberries is similar to the frost that forms on the inside of your windows on a cold day. In the package of raspberries, the condensed frost cannot get back into the food, so as the frost builds up, the berries lose more and more water, leading to freezer burn.

To make the problem even worse, modern freezers are frost-free. Defrosting a freezer used to be a major undertaking. Frost that built up on the walls of the freezer had to either be chopped away or melted by turning the freezer off, usually after the frozen food within had been transferred to someplace else to be kept frozen. Modern frost-free freezers use a refrigeration cycle that causes the temperature within the freezer to climb above the freezing point for a short period. Temperatures of 50°F have been measured inside a frost-free freezer! These temperatures are good for getting rid of frost on the walls of the freezer, but play havoc with frozen foods.

At the elevated temperatures during a frost-free cycle, some of the ice in the raspberries melts and this additional water readily migrates into the warmer package headspace. The result is substantial moisture loss out of the berries. When the temperature cycles back down again that water refreezes and a layer of frost is formed around the berries.

What's the solution to freezer burn? Maintaining your freezer at very low temperatures without a frost-free cycle can help, but packaging is the key. A package that has a good water vapor barrier is essential, as it eliminates any headspace between the package and the food where possible.

Vacuum packaging can remove most of the air in the headspace and prevent formation of berry ice cubes or discolored meat. Just be careful with those vacuum packers that are strong enough to crush a soda can, unless you want crushed raspberries.

20

Lucky Charms – A Lesson in Creativity and Marketing*

How do you invent a new cereal? In the case of Lucky Charms, it was a matter of putting two existing products together and recognizing that the combination was something unique – and tasty. Maybe even magically delicious.

As the story goes, in 1963 John Holahan was part of a team experimenting with new cereal ideas at General Mills. He decided to cut up Circus Peanuts into a bowl of Cheerios, sort of like cutting a banana into your cereal. Sprinkling the orange-colored, banana-flavored, crusty-textured marshmallow Circus Peanuts onto his cereal was an inspiration to Holahan, who knew this sweet cereal would be a hit with kids.

The cereal scientists at General Mills then worked with the marshmallow scientists at Kraft, where Circus Peanuts were made at the time, to develop a cereal marshmallow with the right properties. Although Circus Peanuts may seem pretty dry and hard, especially after sitting for a while in an open bag, they still have much more water than most cereal pieces. In order to make a stable cereal, one that could last for months on the shelf, they had to find a marshmallow with really, really low water content.

Moisture migration is a problem that occurs in many foods – take Raisin Bran, for example. Even though raisins are dried grapes, they still have much higher water content (or activity) than the cereal flakes. Over time, the water migrates from the raisin into the cereal flake, causing the raisin to dry out further and the cereal to pick up moisture. The cereal doesn't change noticeably because of

---

* Not published as a column in *The Capital Times*

the limited amount of water that migrates compared to the much larger mass of flakes. However, the raisins lose enough moisture to turn into hard, tooth-breaking nuggets, at which point, most of us toss the cereal because it's not worth gnawing on the leathery raisins.

The General Mills and Kraft scientists were worried about the same thing happening, with marshmallows drying out in their new cereal. So, they came up with a new method of making marshmallows. Using an extruder, a rope of marshmallow with the right shape was forced out of the die and cut into little pieces. The marshmallow bits were dried slowly until they had about the same water activity as the cereal, so neither marshmallow nor cereal piece changed during storage.

The extrusion process also allowed them to form shapes with different colors. The dried marshmallows, called marbits, provided a unique contrast in both color and texture to the cereal bits in Lucky Charms. The original marbit was a multi-colored rainbow, but it was not long before new shapes and colors were born. Pink hearts, purple horseshoes, blue moons, and of course, green clovers to go along with Lucky the Leprechaun, the mascot of Lucky Charms.

Even though Holahan knew he had a hit cereal, there still remained the problem of finding a name that would sell it. The marketing people suggested developing the cereal around the concept of charm bracelets, a fad that was popular in the 1950s and 1960s (unfortunately, charm bracelets fell victim to disco pendants in the 1970s). Somehow charm bracelets turned into a Leprechaun, but the concept of Lucky Charms was retained. From a marketing standpoint, having kids chase after the Leprechaun's Lucky Charms has been a winning combination.

But what has allowed Lucky Charms to remain such a popular cereal aisle stalwart? Whereas other cereal brands, like Cheerios, have focused on consistency and lack of change over the years, Lucky Charms has thrived on continuous change. General Mills is always experimenting with marbits with new shapes and brighter colors. From pots of gold to a Leprechaun hat with a clover leaf inside, new marbit ideas are continually being introduced. This constant change is part of what keeps Lucky Charms near the top of the cereal aisle, with sales jumps matching the introduction of each new marbit shape.

What was initially a fortuitous combination of two successful products has turned into a classic cereal. Lucky Charms is a kid's favorite, but has a cult following among adults. There is even a test of sexual personality based on your marbit preference. Like the pink hearts? You're a romantic type. But, look out for those who like the purple horseshoes – they're into kinky things, like chocolate pudding, but that is another column.

21

Developing New Ice cream Flavors

Chunky Monkey, Cherry Garcia, Chubby Hubby, and Phish Food, where do Ben & Jerry's come up with ideas for its new flavors? Ice cream scientists at Ben & Jerry's labs spend their workdays dreaming up and developing new flavors to tickle our fancy. It's very competitive in the ice cream aisle and they work hard to continually come up with ideas to attract our business.

Creative thinking is needed to develop a new ice cream flavor. It requires finding ingredient combinations, and names, that are unique and grab our attention. Most importantly, they have to taste good. But sometimes it's more than that. One of the most interesting Ben & Jerry's ideas involved "thinking outside the pint." For those who think a pint is too large for a single serving, it's a smaller-sized container (3.6 oz) that comes with its own spoon.

Hmm, do not they call those Dixie cups? Sure, but marketing is also an important part of new products – it's all in how they're marketed. Some marketing people think they can sell any product, no matter how horrible, with the right advertising pitch. Maybe, but marketing is easier when a product "sells itself," like some of the new flavors developed by Ben & Jerry.

Let's take a look inside an ice cream development lab to get a sense of what it takes to develop a new flavor. Let's follow a team of young ice cream enthusiasts as they develop new flavors. After some initial probing about likes and dislikes, the kids hit on two flavors of their own design. One, called Red White You're Blue, is vanilla ice cream with strawberry marshmallow swirls. You add the blueberries yourself for a patriotic treat. The second ice cream, chocolate with swirls of fudge and caramel, is called Mud and Mudder.

Where do new product ideas come from? The initial germ of an idea may come from someone seeing things in a unique way, as with strawberry marshmallow swirl. Our group of ice cream lovers developed an ice cream flavor that they liked and then brainstormed to find a creative name.

Sometimes ideas come from the strangest places and at the strangest times. You may see or hear or smell something that resonates in your mind, and know that a new product would be good. Sometimes 'new' ideas are just the old ideas with a new twist, like the new Ben & Jerry's mini-sizes (like Dixie cups). The key is to be open to such new ideas and to continually look for interesting ways to put things together.

When developing new flavors, there is a lot of experimenting with different possibilities, but all of our trials are based on scientific principles. We need to understand the science behind ice cream so that we maintain all the desired elements (ice crystals, air cells, etc.) that make a premium product. We also have to work within the constraints of a commercial manufacturing system – we have to be able to make the product consistently, efficiently, and cost-effectively.

Once you have hit on what you think is the next big winner, how do you evaluate its success in the market? Most companies rely on consumer testing before rolling out a new product. Just because experts in the company think highly of a product doesn't mean consumers will buy it. Consumer testing is an important part of putting out a successful new product.

Often, companies will either bring in a group of people in the target market to evaluate the new product or they will take the product to someplace like a mall where average consumers can taste and evaluate the new product. With the right consumer input, they can tell whether the product is ready for market or needs to go back into the lab for further developments. If they've gotten it right, the consumer test panel will be a big success.

Ben & Jerry's made it big by creating new and unique flavors of high-quality premium ice cream. Their most popular flavor is – can you guess? It's Cherry Garcia, a delicious ice cream with lots of cherries, named after a well-recognized rock star.

What flavor of ice cream would you develop? Why do you think it would be a national best-seller?

22

Oreos Spawn Host
of New Products*

Oreos are America's top selling cookie, favorites for almost 100 years. They have also spawned what must be a record number of new products and a host of interesting recipes. We'll explore how new products are derived from current icons, but first, some history.

The Oreo was first developed in 1912 as a product of the National Biscuit Company (formed in 1898 by a merger of three large biscuit companies), which later became Nabisco, which is now owned by Kraft Foods. Interestingly, the Hydrox cookie, which some may consider a cheap knock-off of the Oreo, actually pre-dated the Oreo, coming out in 1908. Hydrox was a product of the Sunshine Biscuit Company, a company formed in 1902 by a couple of former employees of the National Biscuit Company. How's that for gratitude?

Although the name Oreo is now a household and cultural icon, there is some uncertainty about where the name, Oreo, actually comes from. One version is that the center of cream, 're,' goes between two chocolate "o's," or cookies. So, 'o'-'re'-'o' spells Oreo. Whatever the origin, they are now technically known as OREO Chocolate Sandwich Cookies.

Oreos are a true food icon, with one estimate saying over 7.5 billion Oreos are eaten each year, or over 20 million Oreos per day. That's a lot of Oreos.

The Oreo has changed little over the years – why mess with a success, or don't fix it if it isn't broken. However, one recent change is worth mentioning.

---

* Not published as a column in *The Capital Times*

The original Oreo was made with a shortening from partially hydrogenated soybean oil, which as we all know is a source of *trans* fatty acids (the original had 2.5 g of *trans* fat/3 cookie serving). Due to recent health concerns with *trans* fatty acids (see Chapter 16), most food manufacturers have spent considerable effort replacing partially hydrogenated fats in their products. Kraft is no exception, but getting exactly the same crunchy cookie and creamy filling without the *trans* fats is critical to a food icon like the Oreo.

Rumor has it that Kraft tried 250 different formulations before settling on a mixture of palm oil and canola oil to replace the soybean oil. The new version, with zero *trans* fats, is tough to tell from the original. However, as with many food manufacturers, Kraft is replacing the *trans* fats with saturated fats, also considered to be bad for the heart. It may be the price of enjoying an Oreo cookie.

The Oreo is an excellent example of how a popular brand name has been used to broaden the market. From Double Stuffed Oreos to Uh-Oh Oreos, numerous brand extensions have appeared over the years in an attempt to continually expand the market. Following the recent trend of miniaturizing foods to help us maintain a balanced diet, Oreo minis are now available in 100 calorie packs. These new products are designed to continually spark interest and increase sales.

However, such line extensions, as new products that play off an existing brand are called, often come at a price – decreased sales of the original product. Food manufacturers carefully consider the overall balance of a product portfolio when bringing out off-shoots of popular brands. Companies are not willing to market a new product that will sacrifice sales of the original brand.

Oreos are also a great example of new products based on the original concept of cream-filled chocolate cookies. Products like Oreo Cookie ice cream are called 'flankers' because they bring the popular brand name and market identity to a whole new product category. Oreo O's cereal and pie crusts are other examples of flankers spawned by the original Oreo cookie. Oreo-based flankers are new products whose appeal is in part due to our affection for anything Oreo.

As with many popular food icons, people have experimented with various wacky ways to enjoy them. This is certainly true of Oreos. What's the wackiest way of eating Oreos that you've tried?

How about deep-fried Oreos? Dip the intact cookies in some pancake batter and deep fry until golden brown. Another wacky Oreo product is Oreo Dirt Cake – it tastes delicious and it even looks interesting. This is a kid's sand pail filled with layers of crumbled Oreo "dirt" and a cream cheese pudding, and decorated with Gummi worms.

23

Sparkler Spice! for Your Veggies?

What does it take to start your own business? It takes a good idea, a new product or a novel application, and a lot of money. But even that's not enough, as many would-be entrepreneurs who have found out the hard way will tell you. It takes a lot of guts to quit your day job and plunge into the uncertainty of a new business venture. It also takes eternal optimism, even through the darkest day's that you will succeed.

Lynn Hesson, the president of Raven Manufacturing, quit his day job several years ago when he decided to push forward with his idea to manufacture and sell Exploding Pops!. His product is a lot like the more well-known Pop Rocks, a product originally developed at General Foods and now licensed to a Spanish manufacturer. Hesson's idea was that a US-based company making a similar product might have an advantage over an international supplier. He also had lots of ideas, some pretty wacky, of how he could distinguish his product from its competitors.

How do you start a business, especially when you're a lawyer and know very little about making candy? Hesson tapped into some local Food Science expertise to provide a sounding board and technical assistance to get a manufacturing system up and running.

To make Exploding Pops!, a sugar syrup is boiled to over 300°F, then high-pressure carbon dioxide (600 pounds per square inch) is injected under agitation to make small bubbles. The molten sugar mass is cooled, while still under pressure, in large torpedo-shaped tubes. The quickly cooled sugar mass solidifies into a sugar glass, freezing the small carbon dioxide bubbles into the glass matrix. When moisture dissolves the matrix, like what happens when saliva hits the powder in your mouth, the walls holding the bubbles can no longer withstand the internal pressure and POW! – the bubble explodes. As Raven says, it is a party in your mouth!

With lots of help from various experts, in both food and equipment, the technical aspects of constructing a plant to make Exploding Pops! became a reality. However, because of the difficulty in mastering the technology, it took two years from the idea stage to manufacturing the first product. That's a long time to hold onto a dream – and the financial backing needed to pull it off. Small business loans are available to help people start a business, but Hesson eventually had to use personal resources (to wife Julie's dismay) to carry him through the long start-up time.

Even before the product was being manufactured, Hesson was beating the bushes for contracts with food manufacturing companies. Since his idea was to sell Exploding Pops! as an ingredient in other foods, from cereal to ice cream, Hesson had to convince companies to give his product a try-out. Not only did he have to develop the manufacturing site with little help, but he also had to be the technical sales person for the company to give his product visibility.

Even more, Hesson had to be the main product development person within the company. From sugar-free variations to heat-resistant popping candies, he spent many hours working out formulations.

A recent idea to distinguish Raven's product is Sparkler Spice! Exploding Pops! are mixed with savory flavors to make a powder that can be sprinkled on your veggies. You can get a flavored (butter, barbecue or cheese) powder containing Exploding Pops! that, when put on your hot corn, pops and fizzles in your mouth. Hesson claims it's a way to get your kids to eat their veggies, and they can have fun playing with their food at the same time.

After a few lean years, with off and on orders, and an investment in some packaging equipment, Hesson's business looks like it may finally turn the corner into profitability. It has taken a lot of work, a lot of anxious moments, and a lot of fortitude. Eternal optimism is also mandatory.

So for those of us who can get through the anxious moments and succeed, the rewards of starting your own business are enormous. Watching something you've built from just an idea blossom into a self-sustaining venture is gratifying. And there's no way you can complain about the boss!

24

It Is All in the Packaging

Darn, I spilled that big bag of M&M's all over the floor again. I was trying to open the bag gently, but it wouldn't give, until schwapp, the entire bag ripped apart. Sometimes it seems like food companies go out of their way to make it difficult to open their packages. How about tearing the top of a box of crackers or cereal and then not being able to close it again? Why do we need all that packaging anyway?

Food companies spend considerable time in developing the best package for their foods. It's not just our convenience that's important; the package must serve multiple purposes.

First and foremost, the package must protect the food. It must keep out nasty stuff like dirt, microbes, insects, and saboteurs. Nothing is more disgusting than finding a bug in your cereal, so the package must keep critters out. And tamper-proof seals guarantee that larger critters haven't opened the package before we buy it. The package must also be able to withstand the rigors of transportation from the manufacturing plant to the grocery store, keeping our food intact. Some food companies have entire labs dedicated to package testing, from vibrators that simulate trucks driving on a bumpy highway to machines that drop the package from the height of a fork lift to measure how much force it can withstand.

Second, the package must educate the consumer. Required information, mandated by the government, must appear on the package. Nutritional labeling, the serving size, and any caveats about eating the food (such as "may contain peanuts") are examples of information that must appear prominently on the package.

Third, the package should "sell" the product. The package is part of the sales pitch, enticing us to eat the food within. There are regulations on what claims can be made or how tempting the food can look on the package, but marketing people are experts at making food sound appealing. The pictures, or recommended servings, have to at least be realistic, but there's no guarantee that you or I could make the product look that good. My macaroni and cheese never looks half as appealing as the picture on the package. Must be all in the presentation.

Fourth, the package needs to be as environmentally friendly as possible while still meeting the first three requirements. This is probably the aspect of packaging that consumers complain most about. Opening an outside package just to get to another interior package may seem excessive, but if put in the context of protecting the food, it makes more sense. Double wrapping, a necessity on foods from hard candy to cereals, preserves the food even after the outside package has been opened.

Look at Jolly Ranchers – they are individually wrapped and sold in an over-wrap bag. Once the primary bag is opened, the individual candies are still protected from the humidity, preventing that stick-to-the-wrapper fuzz that forms on the surface when they pick up moisture from the air. OK, the wrapper at least slows down the sticky development. Unlike Oreos left to get soggy in an open package, the double-wrapping of Jolly Ranchers helps protect them and enhances consumer enjoyment.

Recent advances in packaging have made things much easier for many consumers. For example, the development of resealable packages for everything from cheese to meats has been a real break-through. We get the convenience of using only a portion of the food in the container, but can preserve the rest for later use while mini-mizing concerns about subsequent deterioration.

Why don't all products come in resealable bags? One pound bags of M&M's® would be an excellent choice for resealable bags, but this brings us to the final concern of the package – the cost. Ideally, the cost of the package should not be so high that it affects the cost of the product. In reality, however, the package may sometimes cost more than the food inside. A 12-ounce aluminum soda can is probably worth more than the carbonated sugar syrup inside.

The next time, you're struggling to open a package or think a food's packaging system is excessive, think about it from the manufacturer's standpoint. Does the package serve all the purposes it should?

25

Shelf Life Dating – Good or Bad?

Like most of us, you root around in the dairy cabinet to make sure you get the milk carton with the latest date, figuring that the later the date, the better. But, are you sure that's the carton with the freshest milk? Milk cartons are not like packages of batteries, where you can squeeze the two electrodes on the package to see how much charge is left. You can't tell if the milk is still good until you open it, but most grocery stores frown on that.

Sometimes, even when we get the carton with the latest date, we get milk that goes bad before it reaches that date. Why is that? A shelf life date that's not reliable is worse than no code date at all.

Those shelf life dates are based on ideal storage temperatures, which for milk would be in the refrigerator at 45°F or less. Unfortunately, our milk is not always stored properly. What happens to the milk between the cow and the store determines how much shelf life is left, particularly if the milk stays out at warmer temperatures for any length of time.

For example, a milk crate may have been left sitting out on the loading dock after being taken off the delivery truck. Who knows, perhaps the stock clerks went on break just when those cartons were unloaded. There they sat, at whatever temperature it was outside, until someone finally put them in the refrigerator. Your mother told you not to leave the milk out on the counter for a good reason – room temperature promotes microbial growth. The longer the carton sits out, the more microbes grow and the less shelf life the milk has remaining even though the date may say it's still good. No matter what the shelf life date says, if you leave milk out on the counter it will not last very long.

Wouldn't it be great if there was a little device on the side of a milk container, similar to the strip on a battery pack that allowed us to see how much shelf life was left in that particular carton? We could buy milk based on its actual freshness, rather than on some average shelf life date based on the conditions milk is supposed to experience.

Actually, devices that can be related to true freshness of food products, particularly those normally found in refrigerators and freezers, do indeed exist. In the food industry, these devices are known as TTIs, or temperature–time integrators, because they show the accumulated changes in storage temperature. TTIs are color-coded strips or circles that change color at the same rate that the food goes bad. If the milk stays in the refrigerator, the TTI doesn't change color very fast, but slowly turns over the two-week period of shelf life. But, if the milk sits out at room temperature, the color changes rapidly indicating that the milk is spoiling. Just by looking at the color on the TTI, we can tell which container is still the freshest. It might not be the one with the latest freshness date, but the one that was stored at the best temperatures.

If these devices exist, why don't we use them? As is often the case, it's primarily a business decision. Most studies have proven the validity of TTIs, but they are not infallible. Thus, companies are reluctant to put TTIs on their products since it means consumers might be tossing milk that's still good. More important, retailers might have to toss milk that is still fine to drink. Besides, the extra cost of the TTI raises the price of the milk.

Packaging scientists continue to work on improving the reliability of these devices and it's likely that, in the future, we will see measurement strips on cartons of milk and ice cream, similar to what we now find on battery cases. When the color is wrong – meaning no shelf life left – toss it out.

26

Intelligent Packages*

Wow, a widget helps to put foam on a can of beer when it's opened. What will they think of next?

Perhaps a package that tells you when the fruit inside is ripe and ready to eat? That would take the guesswork out of squeezing pears to see if they're ripe.

What can you imagine the food package of the future will do?

Traditionally, food packages serve to contain and protect the food from the environment, as well as being a billboard for promotion of the product. However, food packages are rapidly taking on novel tasks. Developing packages that enhance either safety, shelf life or convenience of foods is a hot area these days, with numerous active or intelligent packaging ideas being commercialized.

Active packaging may be defined as a package that does something to enhance its performance, like removing or adding components to extend shelf life. The beer can widget counts as active packaging – it enhances performance by creating beer foam. A package that contains some type of indicator to provide information about aspects of the history of the package and/or the quality of the food would be an intelligent package. A ripeness-indicating fruit package fits that category.

One type of active packaging material, if you can call something active when it just sits there, is oxygen-scavenging plastic. An oxygen-scavenging layer is embedded between two plastic layers. Think of a plastic film sandwich – a layer of oxygen-scavenging polymer film contained between plastic top and

---

* Not published as a column in *The Capital Times*

bottom layers. If any of the oxygen molecules try to get in from the outside, they get gobbled up by this active package layer. One manufacturer claims that using this technology extends the useful shelf life of sliced turkey lunch meat to 55 days in the refrigerator. Keeping oxygen away from the meat prevents oxidative degradation of the meat and inhibits microbial growth and color change.

Other examples of adsorbents used in active packaging include carbon dioxide, ethylene (a ripening agent), moisture, and even flavors/odors. For example, the little sachets in dried beef jerky packages are there to soak up any moisture that gets through the package and prevent premature deterioration of the beef.

Another type of active package releases anti-microbial compounds, like nisin or sorbic acid, during storage to prevent spoilage of foods like meat and cheese.

One type of intelligent package is the self-heating can. Although not a new concept, having been first developed around the 1900 s, some recent developments have renewed interest in cans that can be used to heat foods in self-contained cans. When water contained in the bottom of the can is released into quicklime (or calcined limestone), they react to form calcium hydroxide. In this reaction, a substantial amount of heat is given off, which goes into heating the contents of the can. A can of hot coffee, cocoa, tea or soup is ready in minutes with this technology.

Time–temperature integrators (see Chapter 25) are another form of intelligent packaging that has been available for many years although they have seen limited applications.

Microwave doneness indicators react to the combination of temperature and humidity (steam) as a food is cooking in the microwave. They can be used to indicate when a microwaved food is done and safe to eat. Such a system might take the guesswork out of popping microwave popcorn.

And if these developments are not enough, consider these future possibilities. How about printing low-cost transistors and/or antennas onto a package to broadcast images and sound. With this technology, it's possible for the manufacturer to educate the consumer, for example, about the nutritional value of their product, or more.

How about a container of milk whose package senses that it has warmed up too much and then has the ability to speak to anyone who will listen, begging to be put back into the refrigerator. Now, it would be both intelligent and active if it could actually put itself back into the fridge.

27

Juice Boxes for Your Convenience

Juice boxes are a great example of how new developments in the food industry can make our lives simpler. Before about 1980, juice was pretty much sold in multiple-use glass bottles or plastic containers. Now, much of our juice is sold in single-serving juice boxes or pouches, with your own straw included.

The juice box had its start in Sweden during World War II. The inventor of the juice box, Dr. Ruben Rausing, was working on top-secret coated papers that would hold liquids without leaking. By 1951, he had patented a design for a juice container, in a tetrahedron shape, and in 1952, the first juice-filled packages appeared. His company, Tetra Pak, named after the shape of his first container, has been the leader in developing convenient food packages ever since.

It wasn't until the 1960 s that the box shape, as we know it, was developed and it wasn't until the late 1970 s that juice boxes made their appearance in the US. By then, Tetra Pak was making more than 20 billion packages each year.

The juice box, known as the Tetra Brik, actually begins life as a huge roll of packaging material, in this case a multi-layer laminate of paper, aluminum foil, and plastic. The packaging material is unrolled from a spool and the box shape formed just as the juice is filled into it. A machine quickly cuts the sheet of package into appropriate lengths, overlaps flaps to make a seal at the bottom, fills the enclosure with juice, and then overlaps the top flaps to close the container. The final step is the attachment of the little plastic-wrapped straw to the side of the box.

To make sure the product is safe to drink, the juice is pasteurized (heat treated) before it is filled into the box, which itself has been sterilized with something like a light spray of hydrogen peroxide. Pasteurized juice and sterilized box meet in an aseptic (germ-free) environment to ensure a product with long shelf life, safe from microbial growth.

Juice pouches are even easier to form, fill, and seal. In this case, a roll of the packaging material (a laminated foil with plastic) unwinds into a machine that folds the sheet in half, cuts the right length, seals the bottom and side, fills the newly formed pouch with juice, and seals the top. Again, the plastic-wrapped straw is slapped on the side in the final step. All of these take place so quickly that it's a blur to the human eye.

Let's look at what's inside those juice boxes and pouches. No matter what flavor juice box you buy, the main juices used are typically either pear, grape or apple juice. Desired flavors are added through minor ingredients, usually a little bit of a juice concentrate, like cherry juice, and artificial flavors. Why? Apple, grape, and pear juices are the least expensive juices available, so they make up the bulk of the juice. Other, more expensive, juices are used more sparingly, for flavoring only.

Juice boxes are now used for much more than juice. A variety of drinks can be found in "juice boxes," including shelf-stable milk, soy milk, sport and fitness drinks, broths, specialty teas, and even wine. Last year, a large winery marketed Sangria in a juice box. That would be nice in your lunch box.

Last year, Tetra Pak produced 105 billion packages, about 15 per person for everyone on the planet. Where do all those packages go after the juice is gone? Although Tetra Pak claims their packages are recyclable, most juice boxes go into landfills. Juice boxes are a convenient single-serving container, but their effects on our environment are yet to be fully appreciated.

28

Beware of Low-Carb Diets

During the heydays of low-carb diets, there were low-carb versions of nearly every food, from chocolate to ice cream.

Regular chocolate and ice cream are definitely not low-carb products; the sugar, a simple carbohydrate, provides the desirable sweetness and texture. What's a chocoholic or ice cream fanatic on a low-carb diet to do? Replace the sugars with something else. Sugar alcohols, or polyols, are what makes low-carb versions suitable for the sweet tooth on even the strictest low-carb diet.

But, what exactly is a polyol, where do they come from, and why don't they count as carbohydrates? Polyols are derived from sugars through a hydrogenation process and although similar in chemical make-up, they have very different properties, especially in terms of how they're used by the body.

Let's first look at the chemical differences between sugars and polyols. A sugar molecule contains carbon, oxygen, and hydrogen atoms arranged in a particular way. Sucrose, a disaccharide, contains 12 carbon atoms, six oxygen atoms, and 22 hydrogen atoms, whereas glucose, a monosaccharide, has six, three, and 12, respectively.

When a sucrose or glucose molecule is hydrogenated, meaning additional hydrogen atoms are added, they become sugar alcohols, or polyols. After hydrogenation, a sucrose alcohol, called isomalt, now has 24 hydrogen atoms compared to 22 for sucrose (carbon and oxygen numbers do not change). Likewise, a glucose alcohol, called sorbitol, has 14 hydrogen atoms instead of the 12 found in glucose. The extra hydrogen molecules impart significantly different properties than the sugars from which they are derived.

First, hydrogenation of sugars makes them much less digestible in our gastrointestinal system. Instead of 4 calories of energy per gram of sugar eaten, polyols may have as low as 1 or 2 calories per gram. Thus, the "net carbs," or how much is actually utilized in our body, is much lower.

Polyols also induce a lower insulin response than sugars, which make them useful in sugar-free products for diabetics and others concerned with changes in the glycemic index.

Sugar alcohols are also not cariogenic, meaning they don't cause tooth decay and give us cavities. They're used widely in chewing gum for this reason. That's why mom only lets us chew sugar-free gum – they're made with polyols.

Are sugar alcohols alcoholic? No, it's just a chemical term for the specific arrangement of carbon, hydrogen, and oxygen atoms. You can eat these low-carb foods containing sugar alcohols with a good conscience and even drive home after eating low-carb ice cream.

Clearly, sugar alcohols have lots of advantages in our diets; but if they're so good, why haven't we seen more products made with them before now? Why have sugar-free products been primarily specialty products for diabetics?

The answer is because most of them have an intense laxative effect. Sugar alcohols are not very well digested in the stomach and, being small molecules, soak up a lot of water (the osmotic effect). So they not only pass right through, but also pick up water and come out in a hurry. The microorganisms in the intestine also ferment the polyols, leading to gas formation (a problem similar to lactose intolerance). Eat an entire pint of low-carb ice cream covered with some low-carb chocolate and you're likely to get some exercise running to the bathroom.

Believe it or not, there are people who promote a diet based on this laxative effect. Lose 14 pounds in seven days on the Ex-Lax diet (seriously, there is a web site that promotes this approach)! What they don't say is how dangerous such a diet is; the dehydration that accompanies consumption of such foods is a real problem that can even lead to death.

Despite the potential advantages, people on low-carb diets need to be careful not to eat too many foods containing polyols. The negative consequences of eating too much may offset the dietary advantages of low-carb foods.

29

May Contain Peanuts! – What Is a Food Allergy?

"What is food to one man may be fierce poison

to others."

Lucretius

"May contain peanuts" or "Made on equipment that also processes peanuts": these are common statements found on many packages of foods that have no peanuts in the ingredient list. What gives with these statements?

What gives is that the manufacturer is concerned about people with food allergies. Some of us have allergic reactions to certain foods, and peanuts are one of the main culprits. Rounding out the big eight food allergens are milk, egg, tree nuts, fish, shellfish, soy, and wheat.

Food allergies, for the small percentage of us who have them, are serious business. Even a miniscule amount of peanut, or more specifically, the peanut protein, is enough to kill someone who is allergic to peanut. That's right, a person who is severely allergic to peanuts can die within minutes if they eat anything that contains even less than a milligram of peanut protein.

Take, for example, a product like M&M's. Plain M&M's contain no peanuts, yet the package states, "May contain peanuts." Because Plain M&M's are made on the same equipment used to make Peanut M&M's, there's a chance that some peanut residue remains. Even after rigorous cleaning following a batch of Peanut M&M's, tiny bits of peanut may cross-contaminate the Plain M&M's.

The immune system in people with a true food allergy responds to ingestion of these foods (the antigen) by releasing chemicals (antibodies like immunoglobulins) from the white blood cells. When the antibodies react with the antigens, chemical mediators like histamines are released. These mediators induce changes in the body that lead to anaphylactic shock.

Anaphylaxis can be exhibited through a variety of symptoms. The appearance of a rash (hives), flushing of the skin, swelling of the mouth and throat, severe asthma, and weakness associated with a drop in blood pressure followed by collapse and fainting are all potential reactions. An extreme case can even lead to death.

A similar response occurs in some people when they're stung by a bee, take certain medications, or even are exposed to latex.

The term food allergy is often misused, however. For example, people often say they are allergic to milk, when they really are lactose intolerant.

A lactose-intolerant person who drinks a cup of milk has a reaction, but it's not anaphylactic shock. The lactose simply passes through the stomach undigested and, when it arrives in the intestines, is fermented by the intestinal bacteria. Fermentation produces gas (like in beer and champagne), which can be uncomfortable and embarrassing, but it's not quite the same as going into anaphylactic shock. The inability to digest the lactose in milk is called a food intolerance, not a food allergy.

How do food manufacturers deal with the growing awareness of food allergies? Putting "May contain peanuts" on the package is one way, but it's not satisfactory for many companies, and certainly not for people who are allergic to peanuts. The only way to be sure that a food is free from allergens is to make sure that the food never contacts potential food allergens.

Many companies are choosing to design separate processing lines for foods that are completely allergen free – one line for products that might have peanuts and another line dedicated to peanut (or allergen)-free products. Some companies are even dedicating entire factories to process allergen-free products so that there is no possibility of contamination. This ensures that a product is allergen free and does not need a qualifying statement on the label.

The benefit is that people with allergies can eat these foods without worrying. What is one man's food doesn't have to be another man's poison.

30

Uses for Chocolate Pudding*

One of our Food Processing labs involves making chocolate pudding to demonstrate the effects of different starches on pudding taste and texture. Once, after the lab, a student asked if we could make him 100 gallons of pudding. We agreed, not knowing exactly what his plans were for bucket-loads of chocolate pudding. Perhaps, we thought, he has a lot of friends who really, really like pudding?

And why not, pudding is simple and delicious. However, as with many simple foods, they can actually be quite complex. Pudding is complex, in part, because there are several types of products that carry the name. The American Heritage Dictionary defines two types of pudding. The first is a sweet dessert usually containing flour or grain, or some other binder (like blood), which has been boiled, steamed or baked; this is the category that fits the chocolate pudding that we make in lab, although the definition is much broader than that. Pudding can also refer to a sausage-like preparation made with minced meat stuffed into a bag or skin.

You see, puddings are as diverse as Yorkshire pudding, made with bread and roast drippings, and blood sausage, meat held together by coagulated blood. Both are completely different from the chocolate pudding our student wanted, for who knows what.

Pudding, as we know it in America, is a sweet dessert with well, a pudding-like consistency – somewhere between a thick liquid and a soft solid. Starch is the primary component used to thicken

---

* Not published as a column in *The Capital Times*

pudding, but it's not as simple as gravy because protein aggregation also plays a critical role in pudding consistency.

Chocolate pudding can either be the cook-type or instant pudding. Their properties are slightly different, depending on how they're made.

In cook-type chocolate pudding, sugar, corn starch (or flour), milk powder, butter, and cocoa powder are boiled. Upon cooling, the starch and protein (from the milk) set up into a matrix that gives the pudding consistency.

Starch granules contain tightly packed starch molecules, amylose and amylopectin, arranged in a semi-crystalline, onion-layered packing. In cold water, the starch granules generally retain their initial shape as small compact particles. The arrangement of the starch granules determines the consistency of the mixture.

How does corn starch behave in cold water? Try this experiment at home. To a small glass of water, add a small amount of corn starch and mix well. It looks like milk, maybe a little thicker, depending on how much starch you added. Now add a lot more corn starch, to the point where you have about 70 percent corn starch and 30 percent water. How does this mixture behave?

We called this "mind pudding" in the 1960s – can you see why? Roll some of this mixture in your hands. It stays solid as long as you're rolling it. But stop rolling, and it flows. Groovy. Technically, it's a good example of a dilatant, or shear thickening, fluid.

Now, back to cooking chocolate pudding. When heated in enough water, starch granules undergo dramatic changes. As temperature raises, water begins to penetrate into the starch granule, hydrating and expanding each granule. As heating continues, however, the amylose molecules diffuse out of the expanding granule and into the water. Eventually the granule disappears. After cooking starch granules for a minute or so, all of the starch molecules are dispersed in the water.

When the cooked starch cools, the starch molecules undergo gelatinization, where the molecules cross-link into a soft solid. That's what makes gravy thick, as long as you do it right. If you don't get the starch gelatinization right, Thanksgiving dinner gravy is runny.

In pudding, it's more than the starch that provides thickness. The proteins also enhance consistency. During cooking of the pudding mixture, some of the proteins in the milk unfold, or denature, and interact with other components. This adds to the thickness of the starch gelatinization. Pudding consistency is different from gravy, in part, because of protein aggregation.

When we make pudding in the lab, we cook the starch–milk mixture in a jet cooker. It is not a jet engine that cooks it – it's a jet of steam that mixes with the starch slurry and cooks it almost instantaneously. The pudding mixture is injected directly with high-pressure steam, which quickly heats and gelatinizes the starch. The output of the jet cooker is hot pudding, with the starch gelatinized and protein aggregated – just cool before eating.

Compared to cooked pudding, instant pudding is a good example of how food science and technology have made our lives easier by making foods more convenient. Simply stir in milk, let it set in the refrigerator, and eat. No cooking needed.

But that convenience means it's more complex. It's generally got the same type of starch and protein matrix; however, the structure must be developed in a different way. First, instant starch, or starch that has been pre-gelatinized by the manufacturer and simply requires cold water to disperse, is used in the mix. And then milk protein aggregation is done chemically, through addition of salts to the mix that precipitate the milk proteins without heat.

That's why instant puddings don't work with soy milk. Soy proteins aren't aggregated by the salts in the dried pudding mix. You don't get pudding consistency with soy milk.

As you might have suspected, the student's intentions for the pudding were not entirely honorable. He was interested in a cheap supply for pudding wrestling – a popular college activity. Instead of buying cans and cans of pudding, this student was hoping to get a few buckets out of our jet cooker.

What's the weirdest thing you've done with pudding?

31

The Magic of Gelatin

According to the Jell-O Museum website (www.jellomuseum.com), a bowl of Jell-O gelatin "has brain waves identical to those of adult men and women." Does that mean the bowl of Jell-O can think like an adult?

Gummi bears and Peeps also contain gelatin, the unique material that gives these products their jiggly appearance and chewy texture. Maybe they even have the adult brain waves?

What exactly is gelatin and where does it come from? Gelatin is a complex protein obtained by breaking down collagen, the stuff in our body that helps to build healthy, vibrant skin. Animals have collagen too, although they're probably not worried about wrinkles. The typical source of collagen for making gelatin is animal parts, things like skin, connective tissue, and bones. Usually, the collagen for making gelatin comes from cows and pigs, but gelatin can even be derived from fish parts.

To make gelatin, the highly structured collagen in bones and hides is extracted from the rest of the material in the matrix before being partially denatured to yield a protein called gelatin. Actually, the denaturing process is a lot like what happens to pot roast sitting in a crock pot all day. The collagen in the connective tissue contained in the roast, which we would call gristle if we didn't cook the roast well enough, is broken down into gelatin by the moist heat. The result is a soft, tender pot roast since the gristle has been degraded.

Interestingly, if collagen breaks down too much, we get glue.

How much gelatin is in Jell-O? The powdered mix, in a box of Jell-O, contains only a few percent gelatin mixed with sugar, flavor, and color. Add some hot water to allow everything to dissolve and then cool to set it into a gel. Because it's liquid at first, you can pour it into whatever shape you desire. Throw in some vegetables and you have a jiggly salad, or throw in some fruits and you have dessert. There's always room for Jell-O, right?

One of gelatin's unique properties is that it forms a thermoreversible gel. When hot water is added, the gelatin dissolves and becomes liquid. When the liquid cools below its melting point, it turns back into a gel-like structure. Heat it up again and it's a liquid. Cool it again – a gel. Try it at home. Experiment with some Jell-O gelatin, a refrigerator, and a microwave (careful not to heat too much). If you're careful and patient, you should be able to demonstrate gelatin's thermoreversible behavior.

What is a gel anyway? A gel is a unique state of matter. It's not a solid or a liquid, it's somewhere in between. Sometimes these materials are called soft solids. Other examples are yogurt and cheese, both gels of milk proteins.

Actually, a gel is mostly liquid, with only a small portion of the material in solid form. Only a few percent of gelatin is needed to make a semi-solid dessert – it's the gelatin gel that holds it together so it doesn't flow. Most of the Jell-O dessert is the colored and flavored sugar syrup held in place by the gel matrix. That's why it's called a soft solid, one that doesn't hurt when you hit yourself in the head with it.

To make a gel, the gelatin molecules rearrange to form bundles of intermolecular triple helices, which is a fancy way of saying that the individual gelatin molecules bond with each other. This bonding, or cross-linking, of the gelatin molecules traps the liquid and gives gelatin-based products the characteristic jiggly behavior and gummy texture.

Where does gelatin get its brain waves? That's more a matter for Dr. Frankenstein, but Gummi bears or Peeps that talk back – now that would be an interesting eating experience.

32

Pretzels

Break open a pretzel and you'll notice that the outside is deep brown in color, while the inside retains the pale white color of the dough from which it was made. There's an interesting process that causes that brown exterior, but first, some pretzel history.

Pretzels have been around for a very long time, although, as with many foods, their origin is debated.

One source claims that a monk in Italy in the early 600 s developed the pretzel shape while trying to come up with a way to use extra dough from baking bread. Before baking, he rolled the dough into a skinny rope, looped it into a half-knot, and then pressed down on the knot to seal the shape. Supposedly, his intention was to imitate the shape of arms folded across the chest in prayer. The name comes from the Latin word for "little rewards," *Pretiola*.

Hard pretzels contain yeast, flour, sugar, and salt, with other ingredients as desired. After kneading, the dough is first formed into the desired shape, then it's dipped in a special bath and coated with salt before being baked in an oven.

Pretzels may be eaten plain, but many of us prefer to eat them with a condiment, mustard being the most common. As kids growing up in New York, my brothers and I would squirt mustard on top of our pretzels to make them easier to eat. Unfortunately, it seems like a dropped pretzel always lands mustard side down. We lost a lot of pretzels that way.

Pretzels are also enjoyed in a variety of shapes and sizes. Besides the traditional pretzel shape, there are rods, sticks, and even pretzel nubs. We especially liked pretzel rods as kids because we could pretend we were smoking cigars.

Rolling the traditional pretzel shape can be done by hand, but machines with mechanical arms were developed to do this work because hand-rolling pretzels is slow and the repetitive motion hard on people's arms. Supposedly, the fastest pretzel roller could make at the most 40 pretzels an hour. Automated pretzel rollers increased production, but modern technological innovations have increased that even more. Pretzels are now made at a rate of thousands of pretzels per hour in extruders.

The dough is fed into an extruder, which is essentially a cylindrical barrel with a rotating screw that forces the dough through a small hole (the die) at the end of the barrel. The shape of the die determines the shape of the pretzel. After the extruded dough exits the die, it's cut, dipped, salted, and baked.

To tell whether a pretzel is extruded or rolled in the traditional way, simply look at the joints of the knot. Hand-made pretzels have raised joints, where the loop crosses over itself and closes at the ends. Extruded pretzels have no real joints, just a solid mass of pretzel dough. Virtually all hard pretzels sold today are produced by extrusion.

So what gives pretzels their unique brown exterior? They take a very short dip in lye prior to baking. Lye, made of either potassium or sodium hydroxide, is a strong alkali often made by soaking charred wood chips in water.

The lye dip degrades starch in the flour at the pretzel's surface. The broken down starch reacts in the heat of the oven to produce brown colors. The lye dip is too short to affect the pretzel's interior, so when a pretzel is baked in an oven, the surface turns a rich brown while the interior remains white.

What else is the lye used for? It breaks down protein in lutefisk to give the jelly like consistency. And, as the active ingredient in drain cleaners, it breaks down hair and grease in a clogged drain. Yummy!

It is hard to imagine something so potent as lye playing such a beneficial role in making pretzels. But without the lye dip, hard pretzels wouldn't have that brown exterior we've come to expect.

33

Peanut Butter

Do you suffer from arachibutyrophobia? It's a fear related to eating peanut butter, but it's not the fear of peanut butter itself. What's there to be afraid of in this delectable, sticky food?

According to the Peanut Advisory Board in Georgia, peanut butter was first developed by a Saint Louis doctor to provide a nutritious food for patients with bad teeth. They couldn't chew whole nuts, but they could easily eat ground peanuts to get their daily protein. Peanut butter may be the original geriatric food, the start of those all-you-can eat buffet lines that feature soft, easy-to-chew foods.

Natural peanut butter contains ground peanuts and a little salt for seasoning, nothing more. The problem with natural peanut butter, though, is that peanut oil separates from peanut meal and floats to the top of the jar. It takes a lot of elbow grease to mix the oil back in to make peanut butter.

Processed peanut butter, on the other hand, has stabilizers added to hold everything together. The FDA Standard of Identity for peanut butter states that peanut butter must contain at least 90 percent peanuts and not more than 55 percent fat. The remaining ingredients are approved seasonings and a stabilizer, usually partially hydrogenated vegetable oil.

The ingredient deck for typical peanut butter lists "roasted peanuts, sugar, partially hydrogenated vegetable oils (to prevent separation), and salt." Adding sugar makes a sweeter peanut butter, something that many kids like. Salt, a flavor enhancer, brings out the peanut flavor in the same way as it enhances the flavor of roasted peanuts.

The partially hydrogenated vegetable oil is added to keep the peanut oil from floating to the top of the jar. The hydrogenated vegetable fat has a melting point much higher than the peanut oil, so it forms a continuous fat crystal network and holds the liquid oil in place.

Interestingly, the nutritional label also reads that there are zero *trans* fats in the commercial peanut butter even though it contains hydrogenated fats, a known source of *trans* fats. How can that be?

Let's do the math. One serving size is 32 grams of peanut butter. If 3 percent hydrogenated fat is added, there is just less than 1 gram in each serving. However, partially hydrogenated fat contains less than 40 percent *trans* fats, so the total *trans* fatty acid content in one serving is less than half a gram. One-half gram is the cut-off. If there's less than half a gram per serving, FDA allows the manufacturer to put 0 grams of *trans* fats on the label.

What makes peanut butter so sticky? One source says that peanut butter's high-protein content pulls the moisture out of your mouth. That's why a peanut butter sandwich sticks to the roof of your mouth.

That may be true, but a dry turkey sandwich sticks to the roof of your mouth just as bad as a peanut butter sandwich does. A plain cheese sandwich is even worse since there is nothing to provide lubrication.

Another theory about sandwiches sticking to the roof of your mouth has to do with squeezing the air out from between the food and the roof of your mouth, sort of like the vacuum caused by a wetted rubber plug. If this is true, bread, which contains lots of small air cells, would be particularly bad, but peanut butter by itself wouldn't be likely to cause sticking.

No matter what causes a peanut butter sandwich to stick, the good thing is that we can add all sorts of things to prevent sticking. Some people like bananas or even bacon. One sandwich, attributed to Hubert Humphrey, has peanut butter, bologna, Cheddar cheese, lettuce, and mayonnaise on toasted bread with catsup on the side.

What's your favorite peanut butter sandwich combination?

If you haven't figured it out by now, arachibutyrophobia is the fear of peanut butter sticking to the roof of your mouth. But with so many different things to complement peanut butter, from grape jelly to bananas, there's no need to fear the peanut butter sandwich.

34

Cheddarwurst

In Madison, Wisconsin, Brat Fest comes every Labor Day weekend. It is a great way to top off a Wisconsin summer — barbecued cheesy brats washed down with a cold beer. So maybe they aren't all that good for you, but once in a while it's OK to indulge.

Recently, we did a survey of cheese in sausage-type products, sometimes called cheesy brats or Cheddarwurst, or cheese sausage. We were investigating a problem, called cold melt, for the cheese industry. Cold melt is when the cheese gets soft and soggy even though the product hasn't been cooked. The cheese should melt when the Cheddarwurst is heated, not in the package.

If you've noticed the problem, you're one of those people who plays with their food — cuts it up to see what it looks like inside. If you haven't, go ahead and sacrifice a Cheddarwurst to see what's inside. Food manufacturers routinely tear apart their products as a part of their quality control protocol.

What we have found in our random selection of cheese in sausage products was that some of the cheese had already gotten soft and gooey. If you squeezed the sausage, you could make what looked a little like a cheese spread come oozing out of the cross-section where you cut it. Yuck!

To help us figure out what causes this problem, we measured various properties of the sausage and cheese, from water content to fat content. We found that the main culprit was the availability of moisture to migrate, what Food Scientists call water activity. Higher water content generally means higher water activity, but lots of small dissolved molecules, like salts and sugars, interact with water and reduce water activity.

Products where the sausage had much higher water activity than the cheese led to rapid cold melt — gooey, oozing cheese. Some of the water in the sausage migrated into the cheese to balance water activity. The extra water made the cheese soggy; hence, the cold melt.

When the product was first made, the cheese was fine. It's only over time, during the product's shelf life, that water migrates and cold melt occurs. How can a product developer slow down these natural processes and extend shelf life, at least a little?

In this case, there are a couple of approaches. One is to balance the water activity between the cheese and the sausage. No water would migrate; therefore, no cold melt. However, reducing water content of the sausage, or adding more sugar and salt, makes the sausage unpalatable. It may last longer, but it doesn't taste good. And changes to the cheese are limited by federal regulations on what you can call cheese.

The second approach is to put a moisture barrier between the sausage and the cheese, sort of like waterproofing leather boots. In cheesy sausage products, a thin layer of fat or oil would make an excellent moisture barrier. The problem is getting that thin layer to stay on the cheese shreds as they are being mixed in with the ground meat. That's not easy because the fats rub off during mixing.

There is another option, and maybe the best one. Eat the brat soon after its made. Since there's no time for the water to migrate, you can enjoy a cheesy brat with no cold melt — just a cold beer.

35

Ice – From Nature to Frozen Desserts

Ice. It comes in a wondrous variety of forms. Take a close look outside on a cold winter day and you'll be amazed at the variety, from lake ice hard enough to drive on to dirt-encrusted drifts. There's powder for skiing and packed snow for sledding and snowballs. Greasy curbside snow comes from salting and sanding the roads. And, there's even man-made snow on ski slopes – do you know what they use to make that?

Come February or March, you may be sick of ice and snow, but one place where ice is always appreciated, and necessary, is frozen desserts. The refreshing coolness of ice cream and Popsicles is particularly appealing as the weather turns warmer.

The variety of ice in frozen desserts is nearly as broad as that found in nature. Frozen dessert manufacturers choose specific conditions to obtain the desired form of ice for each product. Whether Popsicles, sorbet, sherbet, Italian ices or plain old ice cream, the ice found in each is slightly different depending on what's in it and how it's made.

If you've ever made homemade ice cream or Popsicles, you know that the freezing process is the primary step. And they're different. If ice cream mix was frozen in a Popsicle mold, the result would be neither ice cream nor Popsicle but somewhere in between. That's because the ice crystals formed in Popsicles are a lot different from those formed in ice cream. One important difference is whether it's stirred or not during freezing.

To make ice cream at home, ice cream mix is poured into a metal container, in a bucket filled with ice and salt (brine). The cooling effect from the brine causes some of the water in the mix to freeze on the inside surface of the metal container. A scraping blade in the metal container, turned by an electric motor, cleans that ice layer off the surface and folds it into the middle of the mix where the temperature is still a little warmer. This process is continually repeated as the scraper blades travel around the container and the ice builds up in the ice cream. After, sufficient ice has formed to increase the thickness of the slush inside, the motor shuts off and it's time to enjoy some ice cream.

When ice cream is made correctly, the ice crystals are numerous, small, and fairly uniform (top image), imparting a smooth texture. If the ice crystals get too large, you feel them in your mouth as coarse ice cream.

Popsicles, on the other hand, are frozen quiescently, without stirring (see Chapter 36). Just pour the syrup into the mold and let them freeze. Because of the static freezing, Popsicles contain long, skinny ice crystals (bottom image) and have almost a crumbly texture, as those ice crystals fall away from each other in your mouth.

Clearly, the nature of the ice crystals in the frozen foods plays an important role in the texture and quality. Ice cream with Popsicle-style ice crystals would not be acceptable.

This has led Food Scientists to test freezing aids, similar to those used in making snow on ski slopes, to control ice formation in frozen foods. Many substances behave as ice nucleators, promoting the formation of ice under conditions where it would not readily form. From small silver iodide crystals used in cloud seeding to bacterial cell walls found in nature (a fact discovered at the UW-Madison many years ago), ice nucleators are big business. In fact, many ski areas add a product made of broken-down bacterial cells to the water pumped through the jets of their snow-makers to help promote snow formation.

These same bacterial cell fragments are also being tested in frozen foods to determine if they can help promote ice crystals of the desired sizes and shapes. Imagine simply placing a pouch of ice cream mix, with appropriate nucleators, into your freezer and having it make, without mixing, the numerous small ice crystals needed for good ice cream texture. Scientists are drawing closer to that reality, although challenges still remain.

For example, you might ask if it is OK to eat ice cream made with a bacterial ice nucleator? Sure, the bacteria have been killed and their cells shredded into fragments. They are probably safe, but it is better to avoid eating ski slope snow, just in case.

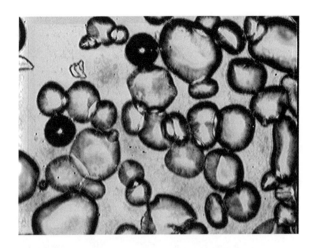

Reprinted with permission: R.W. Hartel, *Crystallization in Foods*, Kluwer Academic/Plenum Publishers, New York (2001).

36

It Is Popsicle Time

Summer is the time to hit the freezer and grab a Popsicle (just don't leave the freezer door open too long, you'll let the cold out!). Frozen sugar (or fruit juice) ices on a stick are easy to eat and help beat the heat.

History has it that the Popsicle was invented (or discovered) by an 11-year old California boy who accidentally left his soda, with a stirring stick still in it, outside on a cold night. The next morning he had frozen soda on a stick. He called it an Epsicle, a play on icicle and his name, Frank Epperson. It took him nearly 20 years to obtain a patent for the Epsicle ice pop, which was later renamed Popsicle by the encouragement of his kids.

Popsicle has come a long way since the days of Frank Epperson. Instead of soda, the ingredient list contains sugars, stabilizers, colors, and flavors. Color and flavor are important for consumer appeal, but it's the choice of sugars and stabilizers that governs the physical attributes of the Popsicle. Without sugars and stabilizers, it would just be a colored ice cube.

Sugars reduce the freezing point of water. That is, sugared water freezes at a lower temperature than pure water, which means that, even though it seems pretty hard, there is still some liquid water in a frozen Popsicle. A Popsicle is essentially a bunch of ice crystals held together by a slush that contains dissolved sugars, colors, and flavors. The more ice (less slush), the harder the Popsicle.

Freezing point depression is a function of the size and the number of sugar molecules. Smaller sugar molecules like fructose and glucose lower the freezing point more than larger sugars like sucrose. That's why the ingredient list of many frozen ice products contains other sweeteners like high fructose corn syrup and fruit juice concentrates. These sweeteners reduce the amount of ice and keep the Popsicle from being too hard.

The other functional ingredients in Popsicles, besides the water and sweeteners, are the stabilizers. These are gums (locust bean gum, xanthan gum, etc.) that provide enhanced viscosity to the liquid before it freezes and then help control ice formation during freezing. The stabilizers also help to prevent melted ice from flowing, giving a more dripless Popsicle.

The manufacturing process is also much more high tech than what Frank Epperson did. Thousands of Popsicles are made every hour in modern continuous automated Popsicle freezers. The sugar concoction is deposited into molds and then submerged in a very cold brine (salt–water) to induce rapid freezing. Sticks are added just as the mixture freezes.

This type of freezing, called quiescent freezing because there is no stirring, results in long, needle-like ice crystals (see Chapter 35). They form initially at the mold surface where it's coldest and then grow radially inwards, toward the center of the mold. All ice crystals lead to the stick. Next time you bite into a Popsicle, check out the pattern of ice formation.

Popsicles have also developed beyond the single stick variety. There are numerous variations to the traditional one-stick Popsicle of Frank Epperson. There are twin pops, with two sticks and two Popsicles, joined at the hip. There's the rocket pop, a multicolored ice pop made by sequentially freezing three or four different layers of sugar syrup. There are Popsicles that glow in the dark, with a glow stick inserted down the middle. You can even make your own frozen ice pop, just like we did as kids, by pouring liquid Jell-O, Kool-Aid or fruit juice into molds and letting them solidify in the freezer.

Next time you pull one out of your freezer, study the ice crystals while you enjoy the cool refreshment. A Popsicle may be a simple treat, but there is still a lot of science involved.

37

Neapolitan Ice cream

Each year, the Ben & Jerry product developers come up with several new flavors. As they entice us to eat more ice cream, they're also entertaining us with their creativity.

Recent new flavors include Vermonty Python, Black and Tan (a take-off on the Irish beer combination of Guinness and Harp), and Neapolitan Dynamite. In a play on the movie, Napoleon Dynamite, they put Cherry Garcia ice cream side-by-side with Chocolate Fudge Brownie, giving something akin to the three-flavored slab of Neapolitan ice cream.

The American version of Neapolitan ice cream, developed in the late 1800s, contains layers of chocolate, vanilla, and strawberry ice creams. Historically, it has its origins in spumoni, an Italian ice cream product originally found in Naples.

Spumoni is a traditional Italian ice cream, often served as two layers of ice cream separated by a layer of fruits and nuts. The ice cream, often chocolate and pistachio, also contains fruits and nuts and may be mixed with whipped cream.

How do you eat Neapolitan ice cream? Do you eat your favorite flavor first or do you save it for last?

An unofficial survey on the web found that about half of the respondents ate their favorite food first, about a third saved it for last, and the rest didn't pick favorites. Perhaps the order in which you eat food, like the flavors of Neapolitan ice cream, says something about your outlook on life?

Do pessimists eat their most favorite flavor first, perhaps worried that someone will take it away or they'll spill it on the floor, or worse? Do optimists, on the other hand, get their least favorite flavors out of the way first so they can really savor the one they like the best?

If so, what flavor do you save for last then? As a kid, I would always eat the vanilla and chocolate first, saving my favorite, strawberry, for last. If this food psychology test has any truth that would make me an optimist, but one with a somewhat quirky flavor choice.

The number one ice cream flavor, in terms of amount purchased, is vanilla. Number two is probably chocolate. Vanilla comes out first because it's the base for other ice cream creations, such as sundaes and milk shakes, not necessarily because it's the most favorite flavor. I would bet that most optimists save the chocolate in Neapolitan ice cream for last.

How is Neapolitan ice cream made? To understand that, we need to understand the ice cream manufacturing process in general.

All ice cream generally goes through a two-stage freezing process. The first freezing step makes a semi-frozen product, similar to soft-serve ice cream, which can be formed and shaped. The second step, called hardening, freezes the product into a hard shape.

To make Neapolitan ice cream, the different ice cream flavors, made separately, are formed into multi-colored slabs in a mold. When hardened, the bricks are cut into half-gallon boxes or ready-to-eat slices.

In a modern ice cream plant, the three flavors of the soft ice cream, made in separate freezers, are fed into a machine that extrudes a slab of three-layered ice cream, which is then cut and sent for hardening.

Where did the Neapolitan Dynamite concept come from at Ben & Jerry's? One of the employees really liked the movie, Napoleon Dynamite, and thought it would be a good concept for a new ice cream flavor. The creative ice cream designers at Ben & Jerry then put together the tri-colored favorite using their existing flavors.

But wait! Perhaps, they can't count too well at Ben & Jerry's, since there are only two flavors in Neapolitan Dynamite. As it turns out, putting three different flavors into a cylindrical pint was too difficult, so they settled for their own unique version of Neapolitan ice cream.

38

Sprinkles or Jimmies?

You might ask for sprinkles. You might ask for jimmies. Either way, you're probably asking for those little chocolate candy pieces that transform plain ice cream into a fun, chewy treat.

What you call these little chocolate bits seems to depend on where you live. In many places on the East Coast, Philadelphia, and Boston, for example, chocolate sprinkles are called jimmies. But, it's not the entire East Coast, since New Yorkers call them sprinkles. And, some people as far west as Michigan and even Wisconsin call them jimmies.

The term sprinkles applies to a wide range of candy-type pieces that are scattered onto ice cream and other treats. From chocolate or rainbow-colored pieces to multi-colored sugar crystals, sprinkles come in numerous sizes, shapes, and colors. Even those little silver or white candy balls, known as nonpareils, qualify as sprinkles.

The term jimmies is thought to have originated in the 1930s in Bethlehem, Pennsylvania at the Just Born candy company. Although most well known for marshmallow Peeps, Just Born also made the small chocolate sprinkles at that time. As the story goes, the man who ran the company's sprinkle-making machine was named Jimmy, and apparently the name stuck.

The Dutch have a similar chocolate sprinkle product, called Hagelslag. They sprinkle chocolate Hagelslag onto buttered bread for breakfast or lunch. No, it's not named after the person who first developed them. In Dutch, Hagel means hail and Hagelslag means pellet.

A hail of small pellets. Exactly, what happens when you pour sprinkles onto ice cream or cupcakes!

Chocolate sprinkles are made largely of sugar and corn starch, with a little fat to soften the texture and some cocoa powder to give it flavor and color. They taste a little like chocolate, but really don't have much flavor of their own. The rainbow-colored sprinkles have no flavor added whatsoever.

The Dutch chocolate Hagelslag, on the other hand, is actually chocolate that's been made into a paste by adding powdered sugar. They actually taste good, whereas chocolate sprinkles need ice cream or cupcakes to be palatable. Does anyone eat sprinkles by themselves?

Other candy products made from sugar and corn starch are candy cigarettes and those little candy dots attached to a strip of paper. Remember them? Although they're much harder than sprinkles — they have lower water content and don't have the fat to soften the texture — they're made in much the same way.

To make chocolate sprinkles, the sugar, corn starch, fat, and cocoa are mixed to form a paste, sort of like a candy Play-Doh. The paste is then extruded in a machine that looks like a pasta press to form thin strands of candy. Multiple ribbons of chocolate candy vermicelli exit the bottom of the extruder.

The candy strands are collected on a vibrating bed and broken into small pieces by shaking. The pieces that are too short or too long are returned to the mixing head to take another trip through the extruder. Those with the right size and shape move on to the polishing stage.

To improve their appearance, sprinkles are coated with confectioner's glaze and wax until they are nice and shiny. Confectioner's glaze is the candy maker's term for edible shellac. A thin layer of shellac puts a shine on the sprinkles just like it does on a nice wooden desk. Confectioner's glaze and wax give the shine to everything from malted milk balls to jelly beans.

Chocolate sprinkles or jimmies, whichever you call them, go well on ice cream, cupcakes, and cookies. But, why stop there? How about a peanut butter and sprinkle sandwich? Or cream cheese and rainbow sprinkles on a bagel? What interesting combinations with sprinkles have you tried?

39

California or Wisconsin Raisins?

Remember those California Raisins® characters from a few years ago? Although they turned into a tremendous marketing ploy, with figurines and even their own television show, their message was simple — raisins are good for you.

Raisins, or dried grapes, are a healthy snack that comes in two varieties — the standard brown raisin or the more colorful and slightly sweeter golden raisin. What's the difference? Do brown raisins come from brown (or red) grapes, in the same way that chocolate milk comes from brown cows?

Nope, you'd be wrong about both. Look at the packages of brown and golden raisins — you may be surprised to see that they both start out as green seedless grapes.

The differences in brown and golden raisins come from two things: how they're made and what else is added. The ingredient list on brown raisins has only one item — raisins. On golden raisins, however, a second ingredient is found — sulfur dioxide. Sulfur dioxide helps to preserve raisins and also impacts the color.

To make regular brown raisins, green grapes are harvested from the vine and generally left to dry in the sun in the rows between the vines. Drying can take up to two weeks, depending on the sunshine and temperature. Southern California has just the right sunny climate needed for drying grapes. That's why the California Raisins® wear sunglasses. If a Wisconsin vineyard tried to dry grapes into raisins, what would the Wisconsin Raisin characters wear; a cheese hat to keep off the rain and snow?

During drying, several chemical reactions take place that lead to the brown color. One important browning reaction in raisins is called the Maillard browning reaction, which was named after a French chemist who first studied the reaction. Certain sugars and proteins react together in a complex series of steps, creating distinctive flavors and brown pigments. This same reaction browns bread during baking and toast during toasting.

A second reaction that leads to browning of raisins is through the enzyme polyphenol oxidase (PPO) contained within the cells. This is the same enzyme that causes apple and potato slices to brown and guacamole to turn brown (see Chapter 13). When PPO is exposed to oxygen, as happens when grapes dry and cells break open, the PPO reacts to form brown color compounds. The combination of Maillard browning and enzymatic browning is responsible for color development in raisins.

Southern California has ideal conditions for browning raisins by sun drying — warm temperatures and intermediate moisture contents. That means a deep tan for California Raisins®.

Where sun-drying isn't feasible, the grapes could also be dried in a forced-air drier. A stream of warm, dry air blows across the grapes to quickly remove moisture. Grapes could be dried in just a few hours in this way. However, the raisins wouldn't be very brown — they would look more like Wisconsin Raisins in winter. The conditions are not right for developing a deep brown color.

But that's exactly what we want when making golden raisins. To make golden raisins, grapes are dried under conditions that don't promote browning so they retain much of the original color of the grape. Raisins that aren't nearly so brown can be made by drying rapidly to inhibit the browning reactions. Adding sulfur dioxide also helps prevent the brown coloring by inhibiting certain steps in the browning reactions.

California Raisins® develop that deep brown color from drying in the sun. Wisconsin Raisins, on the other hand, would be the paler variety because they would have to be dried indoors to avoid the cold, rain, and snow.

40

Eat Your Tomatoes Raw
or Cooked – Just Eat Them

Are tomatoes fruits or vegetables? We eat them on salads, burgers, and sandwiches, along with vegetables like onions and lettuce, but are they really vegetables?

Regardless of what they are, nutritionists tell us to eat lots of tomatoes, in part because they contain lycopene, a plant pigment of the carotenoid family. Once ingested and released into the body, lycopene acts as an antioxidant, scavenging free radicals that can cause harmful reactions and damage cells. The ability of lycopene to scavenge free radicals is likely what makes it active against cardiovascular disease and some cancers.

Nutritionists tell us to eat most fruits and vegetables raw, or with as little processing as possible. However, tomatoes are different. Lycopene occurs naturally in crystalline form in tomatoes and apparently, this form of lycopene isn't readily absorbed from the gut into the body.

To make lycopene easier for the body to absorb, tomatoes need to be broken down and heated. That's exactly what's done in making products like tomato paste and sauce, and their derivatives like spaghetti and pizza sauces. Processing disrupts the cellular structure of the tomato, breaks the bonds between lycopene and other food constituents in the tomato, and promotes absorption of lycopene in the body.

Even tomato ketchup has significantly more available lycopene than the raw tomatoes from which it comes. Wait, is that ketchup or catsup?

Both terms are in use, although ketchup has become the more common term. Heinz has sold tomato ketchup since 1876; however,

Hunt's still markets their product as tomato catsup. Either way you spell it, we use ketchup on a variety of foods from eggs and hash browns to burgers and fries. Both companies promote it as a good source of lycopene. In case you're interested, the world's largest catsup bottle resides in Collinsville, IL, where it was built in 1949 at the site of the Brook's catsup bottling plant.

When we eat ketchup, or any tomato-based product, the lycopene, which is lipophilic (meaning it prefers to be associated with fat instead of water), associates with other lipid-like compounds before being absorbed.

This lipid-like characteristic of lycopene is a reason that some people recommend eating tomato products with some fat. No, not like putting ketchup on burgers and fries, which is heavy on the fat and light on the lycopene. Most nutritionists recommend eating tomato products with low levels (about two teaspoons) of fat to assist in lycopene absorption, like spaghetti sauce with a little olive oil.

The lycopene passes through the digestive tract lining and gets absorbed in the stomach and intestines, where it's eventually distributed to the tissues. After being taken up by the liver, the lycopene is incorporated into the lipoproteins in the bloodstream, primarily in the low-density lipoproteins (LDL). Lycopene accumulates preferentially in certain organs, like the prostate. Perhaps, this is why high levels of processed tomato products are related to a decrease in prostate cancer.

Processing of tomatoes releases the lycopene to make it more available, but unfortunately heating tomatoes is also bad since other vitamins (like vitamin C) are destroyed. Thus, the best plan is to eat a balanced diet that contains both fresh tomatoes and plenty of processed tomato products. For example, spaghetti with lots of tomato sauce accompanied by a salad with fresh tomatoes would make a healthy and delicious meal.

And technically, of course, tomatoes are a fruit because they have seeds. So are green peppers and cucumbers. However, we generally consider them to be vegetables. But does it matter? Eat tomatoes, whether fruit or vegetable, every day, both raw and cooked, to enjoy their health benefits.

41

Fruit Leather

Summer means fresh fruit — sweet, juicy, and delicious. But when there's more fruit than you can eat, how do you put some away for the winter?

A preservation method that's been practiced for centuries with overabundant fruit is drying. Dried berries, apple slices, pineapples, and apricots all make tasty and healthy snacks that can be enjoyed year round. Even dried plums, or prunes, serve a purpose.

Fruit leather is a form of dried fruit. Sometimes called fruit roll-ups, the name implies a certain texture (leathery) and what you can do with them (roll them up).

To make fruit leather, the fruit is first pureed to give a smooth, fluid consistency. Extra sugar can be added, or not, depending on personal taste and the sweetness of the fruit. A little lemon or lime juice might be added to provide some tartness. Adding a little water can thin the puree and make it easier to pour, but adding too much water means it takes longer to dry.

Applesauce is sort of like a fruit puree, one that can be either coarsely ground (with chunks) or finely ground (smooth). It makes an excellent fruit leather when dried.

Although all-natural fruit leathers can be found in the stores, most commercial fruit snacks contain a lot more than fruit and sugar. According to the ingredient list, one common commercial fruit roll-up contains pear concentrate, corn syrup and sugar, hydrogenated fat and emulsifier, organic acids, pectin, flavor, and colors. Regardless of the fruit flavor, pear concentrate is used because of its low cost. And in some cases, there is more sugar than fruit!

Fruit leathers date back centuries; they're thought to have been developed first in the Middle East. Commercial fruit roll-ups, however, are a modern phenomenon, developed in the research labs of General Mills. Food Scientists and engineers developed highly automated methods of producing tons of fruit roll-ups to meet the demands of kids, the primary target audience. The roll-up design, with fruit leather in the form of diamonds instead of squares, is supposedly based on the cardboard tubes at the center of a roll of toilet paper.

Making fruit leather at home is simple. The fruit puree can be poured onto a cookie sheet or a drying tray and then dried in the oven on low heat or left out on a warm sunny day to be sun-dried. Kitchen driers are also handy for making fruit leathers. One summer, I put a cookie tray filled with apricot puree on the ledge behind the back seat of my car, left out in the Colorado sun with the windows cracked a little, to make apricot leather.

Fruit puree is liquid. When poked with a finger, it flows away like any other liquid does. As the fruit dries, though, it gets firmer and firmer, eventually becoming sticky. This sticky point, a well-known phenomenon in sugar-based foods, happens when the fruit mass reaches a critical viscosity. It's the same type of thing that happens when a Jolly Rancher hard candy picks up moisture from the air and forms a sticky surface layer from which the wrapper must be peeled.

After a bit more water is removed, the fruit puree reaches the leathery state – still pliable like leather, but no longer sticky. At this point, the fruit leather is done and can be rolled up in cellophane for later use.

If the fruit mass is dried past the leathery stage, it eventually becomes hard. At very low moisture content, the viscosity is sufficiently high that the mixture reaches a glassy state, similar to that found in window glass (silica) and hard candy (sugar). But, glassy fruit leather is not good because it is too brittle to roll-up.

Finding just the right physical state, between sticky and glassy, is the key to making good fruit leather. However, if your homemade fruit leather does get too brittle to roll-up, simply pretend that you meant to do that and use the hardened fruit chips as a treat on ice cream, yogurt or even breakfast cereals.

42

Preserving Apples for Next Spring

According to the US Apple Association, the old English saying *"Ate an apfel avore gwain to bed, Makes the doctor beg his bread"* turned into "An apple a day keeps the doctor away." But, no one is going to eat apples if they don't look or taste good.

There's nothing better than the crunch of a fresh, crisp apple. Come late spring and summer, though, there's not a fresh apple to be found; only mushy, rotten apples remain. How can we get crisp apples months after the harvest?

Apples, like all fresh fruits and vegetables, go bad with time – they undergo natural respiration reactions after they've been harvested. We actually use ripening to our advantage when harvesting green tomatoes and bananas, since they ripen into red tomatoes and yellow bananas by the time they appear on our grocery store shelves.

But eventually, these ripening reactions, as part of the respiratory process, cause fruits and vegetables to turn bad. Tomatoes get brown and runny, bananas get soft, mushy, and brown, and apples lose their crunch, turning into mealy, mushy fruit that falls apart in your mouth.

Through an understanding of respiration, however, we can preserve the natural characteristics of some produce by judicious choice of storage conditions. Apples are a great example. By simply controlling the atmosphere in storage, the crunch of a fresh-picked apple can be preserved for months.

This technology, called modified- or controlled-atmosphere storage, involves changing the atmosphere around the apple to slow respiration. The first study known on the effects of atmospheric conditions on fruit ripening was done in 1821 when a French scientist showed that fruit in an atmosphere deprived of oxygen didn't ripen as rapidly as in normal air. However, it took over a 100 years for this idea to catch on for commercial application. Currently, one estimate claims that up to half of all apples produced today are stored under controlled conditions to extend shelf life.

Although normal respiratory processes in fruits and vegetables are quite complex and not fully understood, what's known is that oxygen is a necessary reactant. Natural respiration involves oxidative breakdown of organic components, like pectin in the cell wall, to simpler molecules. In this process, oxygen is consumed and carbon dioxide and water vapor are generated as apples turn mushy during storage. As any chemist knows, take away a reactant, like oxygen, or add a product, like carbon dioxide or water vapor, and you can slow down, stop or even reverse a chemical reaction.

That's the principle behind controlled-atmosphere storage of apples: reduce the oxygen content and increase carbon dioxide and water vapor (relative humidity), and ripening reactions slow down. The biochemical processes involved in ripening, which cause crisp apples to get soft and mushy, proceed more slowly in such an atmosphere. By storing apples in a modified atmosphere, along with keeping them at refrigeration temperatures, the shelf life of crisp, fresh-like apples can be extended for up to ten months.

If a little oxygen reduction is good, why not completely remove the oxygen? Unfortunately, when oxygen is completely removed, anaerobic microbial fermentation occurs. Anaerobic microorganisms, those that grow in complete absence of oxygen, cause apples to lose flavor and eventually leads to the production of an alcoholic flavor and skin discoloration. That might be good if we want to make applejack, but it's not the same as a fresh, crisp apple.

In some modified-atmosphere applications, carbon dioxide is used to replace oxygen, at least in part. However, apples are sensitive to carbon dioxide levels above about 5 percent. Higher carbon dioxide levels typically cause roughening and staining of the skin and internal browning, depending on the apple variety. So, nitrogen is usually used to replace oxygen.

Apples to be stored for many months are placed in special refrigerated warehouse rooms. The atmosphere contains about 2–3 percent oxygen, 2–5 percent carbon dioxide, and the rest nitrogen, along with high humidity. The rooms are sealed after the atmosphere has been modified and not opened until it's time to market the apples.

How do you eat apples? Are you an equatorial eater, like most of us, or do you start at the top and work your way to the bottom? Perhaps, you like to cut wedges, peeled or not. But, no matter how you eat your apples, it's now possible to enjoy a crisp apple a day, even in the summer months.

43

Fruitcake: A Scorned Food

Have you heard the joke about the fruitcake recipe? You drink the brandy intended for the cake and toss the rest of the ingredients out the window. How about the joke about regifting a fruitcake and having it return to you years later? Or, the one about the fruitcake that's been the family heirloom since 1892? Fruitcake must be the number one scorned food, a holiday favorite that everyone loves to ridicule.

People are always making fun of it, yet based on the tons of fruitcakes sold each year, someone must be eating it. So how did this negative perception of fruitcake come to be?

Perhaps, the problem comes from the shelf life of store-bought fruitcake. In food processing, one of our goals is to preserve foods for long periods of time to aid in distribution and stock management, but a shelf life of two years (or forever as some people joke) is very long indeed and may be what gives fruitcake its negative image. Twinkies, another scorned food, also have an extremely long shelf life, and the oft-maligned Maraschino cherry has a shelf life of three years!

Fruitcake has a long history, harking back to either Roman times or the Middle Ages, depending on which resource you believe. The idea of putting dried fruit in a bread or cake is nothing new, but the modern fruitcake came into being with the development of advanced preservation methods.

Fruitcake starts with the fruit. If you did nothing to preserve a fruit after picking, it would go bad within a few weeks. The natural respiratory reactions break down the fruit's internal structure leading to softening, browning, and eventually mold growth. It rots if nothing is done to preserve it.

Fruit can be preserved in many ways: by drying, by making it into jam or jelly, by freezing or by canning. But we can also preserve fruit by candying it. The French word is glace', which means to coat with icing or glaze. To glace' (or candy) fruit is to infuse sugar into the interior and then coat it with a sugar syrup. With so much sugar, there's not much that can go wrong with it. No rotting, no mold growth. And it's sweet, the perfect match for cake or sweetened bread.

Candied cherries have a long shelf life. If you candy cherries at home and refrigerate them, they last for about six months. But if you use preservatives with candied cherries, they last for years.

In the same way, fruitcake can have a shelf life from a few months to a few years, depending on the level of preservatives. Fruitcake made at home with no preservatives must be eaten within a few months, whereas many store-bought fruitcakes can last for years. That's where the problem arises. There's a perception that if a food lasts that long, it just cannot be real.

When is a two-year shelf life a good thing? When foods are shipped long distances or stored for long periods, an extended shelf life helps the food manufacturer provide a decent and inexpensive product that doesn't go bad on the shelf. However, if you want the freshest tasting and highest quality product, make fruitcake yourself and eat it within a few weeks.

And if you find a really old fruitcake in the back of your refrigerator, don't worry about eating it, you can always use it as a doorstop or boat anchor.

44

Mom Versus Betty Crocker: Is Cake Made from Scratch Better Than Cake Made from a Box?

June is the most popular month for weddings. Like they say, "Married in June, life will be one long honeymoon." But even if you don't like weddings, at least there's always cake at the end.

Fancy wedding cakes are far different from those boxed mixes in the local grocery store, but does it really matter? Could you replace a mom-made "scratch" cake with a Betty Crocker boxed mix cake instead? We set out to find the answer to that question.

But first, some history. The word "cake" comes from an Old Norse word and, in Medieval Europe, the first cakes were more bread-like, sweetened with honey. Most contained nut and fruit pieces and could last for many months. Cakes were even found in ancient Egyptian tombs.

Modern cakes came about during the mid-1600s with the advent of more reliable ovens, refined sugar, the cake hoop to make cakes round, and, most importantly, a crude frosting. Most of these cakes still contained fruit pieces. Refined flour was not used until the mid-19th century, about the same time butter–cream frosting was invented.

Boxed cake mixes were introduced in the 1940s, far later than other packaged mixes that were introduced during the Industrial Revolution. Initial consumer reaction to these new boxed cake mixes was negative. Traditions conflicted with modern convenience. Mom was still expected to make a homemade cake and the mixes took all the work out of it. But through shrewd marketing, boxed mixes prevailed, even though they produced substandard cakes. Today, cakes from boxed mixes are much better, but do they compare to "scratch" cakes?

Boxed mix cakes definitely have a price advantage. Large companies buy all their ingredients in bulk, paying very low prices compared with consumer's retail costs. To make a chocolate cake from scratch, we need sugar, flour, cocoa, baking soda, eggs, milk, oil, and vanilla. Because we generally have to buy more of each ingredient than the recipe calls for, we end up spending close to $10 for the ingredients to make a cake from scratch. Granted, we might have many of the ingredients on hand in the kitchen already, but we still needed to buy them at some time.

At about $2, a boxed cake mix is a bargain, even if we have to add a couple of eggs and some oil. Plus, the time savings in making the box cake can be substantial. But can we tell the difference in quality?

Being scientifically inclined, we designed an experiment. With the help of a real mom (because it wouldn't be real homemade cake if mom didn't make it), we baked two similar cakes: one from scratch, using a recipe from the back of a cocoa tin, and one from a box mix. We then had the cake experts in a high school philosophy class test the two in a blind taste test. The students decided which one tasted best and which they thought was made from scratch.

Of 22 students, only three thought the box cake was made from scratch. Perhaps not surprisingly, almost everybody could tell the difference between the two cakes. But how did the taste compare?

Despite mom's best efforts, the class was split on which they liked better. Eleven preferred the scratch cake and the other eleven preferred the boxed mix cake. Even mom thought the box cake was pretty good, although she tends to be most critical of her own work.

Which is better — home-made versus boxed mix cakes? The debate continues. Moms will always argue that cakes from scratch taste better, despite the evidence that most of us cannot tell the difference. However, at weddings, nothing will do except the finest home-made cake. It's a special day that deserves the best of everything, especially the cake.

45

Holiday Cookies – Butter, Margarine or Shortening?

Do you make your holiday cookies with butter, margarine or shortening? Or, like me, do you use half of one and half of another?

Experienced cookie-makers know that cookies made with butter, margarine or shortening come out different. Cookies made with butter often spread very thinly, although the buttery taste is nice. Cookies made with shortening hardly spread at all, and have much less buttery flavor. That's why I use a blend of the two – it gives a nice balance between flavor and spread.

The differences in cookies made with butter, margarine or shortening are due to a number of factors, including the amount of water they contain, the types of fat used, and how much fat is crystallized, or solid, at different temperatures.

Other things being equal, more water means thinner cookies. Butter contains about 18 percent water, as does stick margarine, so both cause cookies to spread. Low-fat spreads have even more water and result in cookies that may spread all over the pan. Shortening, on the other hand, contains no water, which is why there is minimal cookie spreading.

The type of fat is also important. Butter comes from churning cream (see Chapter 14) and therefore, contains only milk fat from the cow. Margarine and shortening are made from vegetable oils (cottonseed, soybean, etc.) that have been modified in some way to have desirable physical properties for baking and other uses.

Milk fat is a fairly hard fat. Although only about half of the milk fat in cold butter is actually crystalline that's enough to cause problems when you try to spread it on your bread without letting it warm up a little (which melts some of the crystallized fat).

Margarine, on the other hand, has been designed to have only about 15–20 percent crystallized fat (the rest is liquid), so it can be easily spread on bread without having to be warmed up.

Shortening is made from similar fats as margarine. Crisco, a common household shortening, was first marketed by the Procter & Gamble Company in 1911 to replace lard. Originally named Krispo, P&G eventually settled on a name that was sort of an acronym for the ingredients – crystallized cottonseed oil or CRISCO.

At that time, a process for hardening liquid oils, called hydrogenation, had just been commercialized. By adding hydrogen atoms to unsaturated fatty acids, hydrogenation produces solid fats from liquid oils; like Crisco from cottonseed oil.

The differences in water content and fat crystallinity explain, for the most part, the differences in cookies made with butter, margarine, or shortening. But the temperature at which the fat melts completely is another important factor.

The hydrogenated vegetable fats in margarine and shortening melt completely at a temperature slightly higher than milk fat. Thus, cookies made with butter, with a slightly lower melting point, spread more at baking temperatures.

Then, there are health concerns to consider. There has been much debate about which fat is healthier, a discussion usually related to the saturated fat content. However, nutritionists have recently discovered that certain types of fats, called *trans* fatty acids, are even worse for us than saturated fats (see Chapter 16). Since *trans* fatty acids are typically produced from partial hydrogenation, margarines and shortenings made from partially hydrogenated vegetable oils have come under close scrutiny.

Because of the concerns over *trans* fats, there are now numerous *trans* fat free margarines and shortenings available that are not made with partially hydrogenated oils. And these new *trans* free products undoubtedly behave differently in cookies.

That's an excuse for doing some kitchen science to discover how these new products affect cookie behavior. How do their properties affect cookie spread, browning, and flavor?

46

Animal Crackers or Cookies?

Animal Crackers. Are they really crackers or are they just hard cookies? If they were really crackers, wouldn't you put them in your soup? But do you know of anyone besides Shirley Temple who puts Animal Crackers in their soup (and then sings about it)?

So what's the difference between a cracker and a cookie? Let's look at a couple definitions.

WordReference.com defines a cracker as " a thin crisp wafer made of flour and water with or without leavening and shortening; unsweetened or semisweet." A cookie, on the other hand, is defined as "any of various small flat sweet cakes ('biscuit' is the British term)." What the British call biscuits would encompass both crackers and hard cookies in the US.

Does that clarify Animal Crackers for you? Probably not, so let us look at the ingredient lists for some common Nabisco products to see if we can distinguish cookies from crackers.

Ritz Crackers contain flour, soybean oil, corn syrup, salt, baking soda, and lecithin as an emulsifier. Nilla Wafers, arguably eaten as a cookie and not as a cracker, contain flour, sugar and corn syrup, shortening, eggs, salt, baking soda, and emulsifier. The main difference is the sugar content – cookies typically have more sugar than biscuits, but not always.

Barnum's Animal Crackers are made with flour, sugar (and some corn syrup), hydrogenated vegetable oil, salt, baking soda as a chemical leavening agent, and lecithin. Ingredient-wise, they are more like Nilla Wafers, or cookies, than Ritz Crackers.

However, there is no definition for how much sugar is needed to cross the line from cracker to cookie. In fact, the range of sugar content for both commercial cookies and semi-sweet crackers generally falls between 20 and 30 percent. Cookies also tend to have higher fat content than semi-sweet crackers.

Another difference between cookie and cracker, at least most of the time, has to do with how they are made. Typically, cookies are often stamped (the dough is rolled out and cut with a cookie cutter) or extruded (like in a cookie press). Crackers are often sheeted and layered (or laminated) before being formed and baked.

Laminating involves folding a thin sheet of dough back and forth over itself to make multiple layers. The laminated dough is then formed and cut prior to baking in the oven. To prevent moisture from releasing between layers in the cracker in the heat of the oven, causing huge bubbles or blisters, holes are poked into the dough prior to baking.

Look carefully at Animal Crackers. They all have holes, called dockers, to prevent blistering. Those clever zookeepers even thought to put one docker where the eye should be, to enhance the animal's image.

Animal Crackers have been around for a long time. In 1902, the National Biscuit Company, which later became Nabisco, renamed an existing product called Animal Biscuits to Barnum's Animal Crackers – the product we now buy in the box with the string. The string was added so the box could serve as a Christmas tree ornament.

They say there have been 37 different animal characters since 1902. The last to be added was the koala bear, voted in by popular demand in 2002. No fish or oysters, although you would think they would be the best ones for swimming in soup.

So, are Animal Crackers cookies or crackers? In more than name only, they are truly crackers, but they are one of those products that sort of falls between both the categories. Even though they're crackers, they're too sweet for most of us to put into soup.

Maybe that is why the British call them all biscuits – they avoid all this confusion.

47

Skunky Beer for Oktoberfest?

How long has beer been around? Millennia, at least. The first beer was attributed to the Sumerians about 6,000 years ago. It's said to have been found by accident, perhaps after bread or grain got wet and began to ferment.

After all these years, you would think we'd know everything there was about beer, yet we still are likely to be assaulted by that skunky smell when opening a bottle of beer. Millennia of making, studying, and drinking beer, still have not taught us how to keep skunks from climbing into our bottles.

Sure, we've known for a long time that there's a chemical reaction driven by ultraviolet light that leads to the production of the skunky off-flavor. However, it wasn't too long ago that the details of this chemical reaction were finally deciphered.

Is the chemical produced in beer the same as the chemical that gives the skunk its smell? Not quite, but it is pretty close. The same class of sulfurous chemicals, called thiols, exists in both skunky beer and a skunk's spray. But as Monty Python would ask, is that a striped skunk or a spotted skunk? Didn't know there were different kinds of skunks? And furthermore, they have slightly different chemical compounds that make up their smell.

The skunky odor in beer comes from the same source of chemicals, regardless of whether it's a Spotted Cow, Moose Drool, or Red Stripe. A class of chemicals, called isohumulones, found in the hops, is converted to a thiol when exposed to light. The reaction, which involves very short-lived and very reactive compounds called free radicals, is driven by ultraviolet light, so keeping light away from the beer should prevent skunkiness.

In this recent study, a very sensitive instrument was used to detect the short-lived radicals and help understand the reaction. Technology has caught up with beer drinkers. This understanding could potentially lead to new approaches to preventing skunkiness.

Of course beer makers (and beer drinkers) have known about skunky beer for hundreds of years. One solution to the problem is to drink the beer as soon after bottling as possible so the reaction does not have time to occur. That might have worked years ago, but much of our beer is now made in large centralized breweries with wide distribution systems. This beer needs to last a long time without getting skunky. Thus, we find beer in colored bottles that filter ultraviolet light. Cans work too, but they affect flavor in different ways.

Bottled beer is most often sold in brown or green bottles intended to filter some of the ultraviolet light that causes the reaction. But even that's not enough to completely prevent the reaction. Open a green bottle of beer that's been sitting out in the sun for too long and step away — smells like a skunk died in there. Colored glass only goes so far.

How do some beer makers get by using clear bottles? There are a couple of approaches. First, you can chemically modify the beer during production to remove the components that react with light. No reactants — no skunky beer, no matter how much light you have. This is what at least one of the white-bottled beer brands does.

Another approach is to stuff a lime in the top of the beer to confuse your senses enough that you'll drink it no matter how skunky it really tastes.

48

This Oktoberfest, Drink the Beer, Not the Water

When traveling in certain countries, you often hear "Don't drink the water, drink the beer. It is safer." That's good news during Oktoberfest.

In fact, throughout much of history, it's been safer to drink beer because a source of pure water hasn't always been readily available.

Taking it even further, many beer makers tout the source of their water as one of the reasons for the high quality of their beer. Take Old Milwaukee, for example. No, I mean Coors – the Rocky Mountain water used to brew Coors is cited as one reason it tastes so good.

Is that truth or just marketing hype? Does the water really have an effect on beer quality?

Absolutely, brewing water is an essential ingredient in making a good beer. Beer is over 90 percent water, so you'd expect the water to be important, but it goes way beyond that.

In fact, different beer styles developed in different countries and regions based to some extent on the characteristics of the water found in that region. In order to understand why, we need to know what's in our water.

Water in beer should of course be free of contaminants – we don't want lead or PCBs or pathogenic microorganisms in brewing water. But, other molecules commonly found in water can also profoundly affect beer quality.

In particular, dissolved minerals, leached from the rocks over which water flows, are very important to the brewer. Hard water contains certain dissolved minerals that significantly impact brewing operations.

Boiling water kills bacteria, making both water and beer safe to drink. Boiling hard water leaves a white residue, primarily made of calcium and magnesium salts that precipitate out of the water when heated. Water scientists call this temporary hardness. Even softened water contains dissolved minerals, but different ones that don't precipitate when heated.

The relative composition of the minerals in water is determined by the types of rocks through which the water has passed, which differ from region to region. These minerals impact brewing processes, particularly malting and fermentation. For example, certain minerals have specific effects on the yeast that ferment sugars into alcohol.

Calcium and magnesium, the important components of temporary water hardness, affect brewing in several ways. They give beer a certain "mouth feel." They also affect acidity levels and yeast activity. Even the balance between calcium and magnesium levels is critical to beer taste.

Water also contains dissolved sodium, especially if the water has been softened by the process used in home water softeners. Sodium imparts a salty taste to beer, but too much of it can also have a negative effect on yeast activity.

Trace elements like zinc and copper also impact yeast metabolism, so their levels in water must be controlled for brewing. Carbonates and sulfates impart specific flavors and may affect the extraction of hops, thereby further modifying flavors.

Since these components in water vary widely throughout the world, it's no wonder that beers from different parts of the world taste different. Although today's brewer can artificially control the mineral composition of brewing water, hundreds of years ago, brewers simply used the natural resources at hand.

Thus, beer brewed with the soft water found in Pilsen, in the Czech Republic, produced a typically mild lager, whereas the lagers made with the hard water found in parts of Germany had a much stronger taste.

So feel free to drink beer this Oktoberfest, knowing that the water, while being an important part of beer's quality, is also safe to drink – unless you drink too much and try to drive home afterward.

49

Fresh Orange Juice

Squeeze a fresh orange and you get a delicious drink, a wonderful mixture of sweet and tart. Or not. Sometimes, instead of delicious nectar that brings a smile to your face, your orange yields only a small amount of highly acidic, and not very sweet juice that puckers your lips.

What causes this variability in oranges? Sunshine and rain, the temperature, and maybe even the humidity during the growing season, all affect the quality of plant crops like oranges. These factors also cause the differences among wine vintages (see Chapter 3). The same grapes grown in different years and different regions yield wines of different quality. Perhaps, we need vintage years on orange juice too.

Small differences in sugars, acids, and essential oils or flavoring molecules induced by the differences in weather during the growing season lead to these differences in flavor and taste. Orange juice, or grape juice for that matter, is nothing more than the sum of its chemical components. All that distinguishes sweet from sour juice is the amount of sugar and acid.

Whatever the cause of the variability, food processors must accommodate these differences when they package orange juice for later consumption. When we buy orange juice at the store, we expect to get juice that is always acceptable. In fact, we expect the juice to always taste the same, regardless of when the last hurricane hit Florida. Plus, natural orange juice can only contain certain things; anything else and you can't call it 100 percent orange juice.

How do companies like Minute Maid and Tropicana maintain that level of quality and consistency despite the variability in oranges? If the claim on the label reads 100 percent orange juice, not from concentrate, you know they're only using the components that are normally found in oranges. Nothing else is added. But what options do they have to maintain a consistent quality?

In principle, you could take any orange, measure the amount of each of the important chemicals (sugar, acid, etc.), and then add whatever component was needed to get a good balanced juice. If you got those components from an orange itself, you could still call it 100 percent orange juice.

That's the principle behind standardizing juice. First, separate components that have certain chemical make-ups and then blend the parts as needed to always get the same chemical composition. For example, orange essence (the flavors and aromas) comes from distilling juices or even peels. These essences are often used as perfumes or flavorings, but can also be blended back into juices to provide natural flavoring.

Milk producers use standardization when they skim the fat from milk and then add back just enough cream to make whole milk with a consistent 3.2 percent fat content.

Another approach to standardizing orange juice is to obtain oranges from a variety of sources and regions, analyze each juice for chemical composition, and then blend the juices to get as close to the target composition as possible. This approach requires constant manipulation of ingredients to get a blend closest to the desired standard at the lowest possible cost.

It may seem like a simple job to make orange juice – just squeeze some oranges into a container – but it's not always that simple. Orange juice manufacturers spend a lot more time and energy than we realize to make sure their product always brings a smile to our faces.

50

Apple Cider

One of the sensory pleasures of fall is drinking fresh apple cider pressed that same day at the cider mill. The sweet, crisp flavor of fresh apple cider makes it a delicious and healthy treat. But before you drink raw apple cider, consider this.

The FDA now mandates that all apple cider processors pasteurize their cider to ensure its safety. However, the regulation only applies to manufacturers who sell to retail outlets. Small cider mills may still sell raw cider at their stands, but only as long as there's a warning label attached.

Drinking raw apple cider may be hazardous to your health!

Turn back to the 1990s when several outbreaks of illness associated with *E. coli*, and other microorganisms, were traced back to raw apple cider. Hundreds of people suffered from severe gastrointestinal problems and a few people even died from drinking contaminated cider.

In a study conducted at the University of Maryland, *E. coli* was found on samples of raw apples, on cider mill equipment, in the pressed juice and on the discarded pomace — the residue after pressing. The microbe is clearly present on apples.

Where does the *E. coli* come from? One common theory is that apples picked off the orchard floor are contaminated with microorganisms naturally found in dirt, and even brushing and washing the apples do not remove them all. If so, using only unbruised tree-picked don't should give uncontaminated apple cider, right?

Apparently not, studies have also documented *E. coli* contamination on tree-picked apples, indicating that contamination can occur in other ways as well. For these reasons, the FDA requires that cider be pasteurized.

It's logical to ask though why there's such a concern nowadays. Haven't people been drinking raw cider, and raw milk as well, for centuries? The answer is complicated. One possibility is that microorganisms are evolving in response to new environmental factors. But it's more likely that people have been getting sick for years from drinking raw cider and milk. We just didn't know it.

Until recently, tracing a food poisoning outbreak to a specific food or location has been nearly impossible. Now, modern forensic methods give us better means to work backwards to the source of the problem.

Accepting that cider should now be pasteurized, how can it be done to maintain as much of the fresh-pressed quality as possible?

Traditional pasteurization means heating cider until sufficient numbers of microorganisms have been destroyed to make it safe to drink. University of Wisconsin Food Scientists recommend heating raw cider to at least 155°F and holding for 14 seconds. Others recommend heating to 160°F and holding for at least six seconds. Either way, the number of microorganisms must be reduced by a factor of 100,000 (five orders of magnitude).

However, heating also causes significant changes in quality, particularly when the longer times are used. Heating-induced changes are good when we roast a turkey or toast bread, but in fresh foods like raw milk and cider, heat causes undesirable changes. Pasteurized apple cider is said to taste "cooked" compared to raw cider and has a different color. To many, these changes are unacceptable.

To retain the natural flavor and appearance of raw cider while ensuring its safety, various nonthermal pasteurization methods have been studied. One of the most promising technologies uses ultraviolet light of sufficient intensity to destroy microorganisms. The result is a pasteurized product that retains most of the desirable attributes of fresh-pressed cider. In fact, the FDA now approves the use of ultraviolet light pasteurization for the treatment of apple cider, although cider aficionados still say it tastes different than fresh juice.

This fall, you have a choice. You can drink raw apple cider for the flavor, risking a bout of food poisoning (or worse), or you can drink pasteurized apple cider and be safe. Fortunately, with new pasteurization methods, you can have your fresh cider and drink it too.

51

Egg Nog – A Safe Holiday Tradition

There are as many stories about the origins of egg nog as there are recipes to make it, from nonalcoholic versions for the whole family to those that pack a powerful alcoholic punch. Even though egg nog sometimes contains raw eggs, with proper preparation it can be a safe and tasty drink for the holidays.

Egg nog is most likely derived from an English drink called posset, or spiced milk with wine or ale added. Posset was used as a cold medicine in medieval times. The egg nog we know today is often made with eggs, milk (and/or cream), sugar, and spices, and if desired, your favorite alcoholic beverage.

In colonial North America, rum was added to egg nog to provide the kick. Rum is still the preferred spirit in egg nog in many parts of the country, although it can be made with bourbon, whiskey, brandy, sherry, or nearly any other type of spirit.

Regardless of what spirit, if any, is added, it's still called egg nog. Some say the name egg nog comes from colonial America where rum was called grog, so that egg and grog got shortened to egg nog. Others suggest that the term nog comes from noggin, which can mean either ale or a small wooden mug. A drink made with egg and spirits, served in a small wooden mug might then have been called egg nog.

Egg nog, no matter where the name comes from, has become an American tradition enjoyed by millions each holiday season.

According to a recipe supplied by the American Egg Board, egg nog is made by adding six eggs, a quarter cup of sugar, some salt, one quart of milk, vanilla, and seasonings to taste. The eggs are beaten with the sugar and salt, half the milk is added and the mixture is heated slowly to 160°F. When the mixture attains the proper consistency, so that it's thick enough to coat a metal spoon, it's removed from the heat and the remaining milk is added along with the flavorings. The egg nog is cooled in the refrigerator before serving.

In traditional egg nog recipes, raw eggs are whipped with sugar and milk into a thick foam before cream, spices, and the spirit of choice are added. However, raw eggs are no longer considered a safe food and should be cooked during processing to ensure safety from contamination. Commercial egg nog is always pasteurized to protect against food poisoning.

For many years, the interior of eggs was considered to be almost sterile and eating foods made with raw eggs (Hollandaise sauce, egg nog, etc.) was acceptable. Even though mom might have slapped your hand for stealing raw cookie dough, it was unlikely to cause food poisoning.

However, we now know that approximately one egg in 20,000 may contain *Salmonella enteritidis*, introduced either by transfer through the shell or from within the hen before the shell is even made. The bottom line is that even an egg with a clean, intact shell may still be contaminated. In healthy individuals, *Salmonella* poisoning results in stomach cramps and diarrhea, symptoms that are often misinterpreted as the flu. In people with compromised immune systems, however, *Salmonella* poisoning can be deadly.

In the American Egg Board recipe above, heating the egg mixture slowly to 160°F was sufficient to destroy the *Salmonella* and ensure a beverage safe from contamination. Alternatively, pasteurized eggs, available at the grocery store, can be used and the heating step can be skipped entirely. Or, some people might accept the risk, about five one-thousandths of 1 percent, of contracting food poisoning by eating raw eggs.

Fortunately, the alcohol in fortified egg nog helps protect against *Salmonella* poisoning. Recent laboratory studies show that alcohol kills *Salmonella*, a fact that has been corroborated in studies where the severity of a food poisoning outbreak was correlated with alcohol intake. For people who ate the same contaminated foods, those who drank the most alcohol with the meal were least likely to come down with food poisoning.

Although it's not recommended to drink egg nog made with raw eggs, egg nog fortified with strong spirits can at least reduce the risk of food poisoning.

52

Kool-Aid or Tang?

Are you a Kool-Aid lover? A Tang addict? Although not as prominent today as they were several decades ago, both are still popular ways to sweeten a glass of water.

Common lore has it that the original powdered orange drink, Tang, is an offshoot of the space program. But, both NASA and Kraft Foods, the current manufacturer of Tang, say that's incorrect. Tang was developed and marketed in the late 1950s by General Foods as a modern breakfast drink. It wasn't until the 1960s that NASA took Tang into space, supposedly to mask off-flavors of treated water. General Foods used the space connection as a marketing tool, and Tang has been associated with manned space flights ever since.

Kool-Aid was developed in Nebraska in the 1920s by inventor Edwin Perkins. When he sold the brand to General Foods in 1953, he was making over a million packets per year. Kool-Aid is still the official state drink of Nebraska (19 states, including Wisconsin, of course, claim milk as their official state drink).

What's in Kool-Aid and Tang? Tang contains all the ingredients needed to sweeten, color, and flavor a glass of water, whereas the original Kool-Aid requires added sugar to sweeten. In addition to sugar, Tang contains citric acid, vitamin C, potassium citrate, malic acid, xanthan and cellulose gums, calcium phosphate, colors, and flavors. Kool-Aid contains citric acid, salt, calcium phosphate, colors and flavors, and vitamin C.

Have you ever tasted unsweetened Kool-Aid powder? It's intensely tart! That's because the number one ingredient is citric acid. Unsweetened Kool-Aid powder has to supply the entire pitcher of Kool-Aid with taste, so the ingredients are very, very concentrated.

Because crystalline acids pick up moisture readily from the air, both products contain calcium phosphate (tribasic) to prevent caking. Like rice grains in a salt shaker (see Chapter 18), calcium phosphate crystals keep Tang and Kool-Aid powders free-flowing.

The potassium citrate in Tang is a buffering agent; it moderates the acidity in the drink. Xanthan and cellulose gums are added to provide thickening; that's why a glass of Tang is more viscous than Kool-Aid.

Could you use powdered candies, like Lik-m-Aid and Pixy Stix, to make a sweet soft drink? They contain dextrose, citric acid, colors, and flavors, almost the same as in Kool-Aid and Tang. However, the resulting drink wouldn't be as sweet because the primary sugar in powdered candies is dextrose (also called glucose).

Glucose is also only about 70 percent as sweet as sucrose and at room temperature, only about half as soluble. Powdered glucose is plenty sweet for most kids because it's concentrated in the mouth; but because of its low solubility and low sweetness, mixing it with water would yield something more like a sport drink than Kool-Aid.

Tang and Kool-Aid flavor a glass of water, but don't provide the fizz of a carbonated soda. A 1960s product, Fizzies, was a tablet that was added to water to produce a sweet, bubbly drink. Fizzies were sort of like a good-tasting Alka-Seltzer; the fizz is caused by the reaction of bicarbonates with water.

Originally sweetened with cyclamates, a mix of cyclamate salts and cyclamic acid that is 30 times sweeter than sugar, Fizzies were taken off the market when the government labeled them as carcinogenic. Years later, it was determined that cyclamates really don't lead to cancer, but the damage was done and Fizzies were out of business. But there's good news for Fizzie fans – they're back (www.fizzies.com), now sweetened with acesulfame potassium and sucralose.

Why not sweeten Fizzies with sugar? Turns out that to sweeten a cup of water with a tablet containing sugar would require a tablet the size of a hockey puck. That's too big for most glasses, so a more intense sweetener is needed.

Perhaps, you've heard the rumor that you can use Tang powder in your dishwasher in place of detergent. While it's true that the high levels of citric acid provide a cleaning effect, the manufacturers strongly recommend that Tang only be consumed as a drink and not used as a cleaner – either on Earth or in orbit.

53

Milk Shakes and Brain Freeze

Nothing beats a milk shake on a hot summer day. It's cool and refreshing whether you suck it through a straw or spoon it into your mouth. But, it might also cause sphenopalatineganglioneuralgia if you're not careful.

As with many foods and drinks, what you call a milk shake depends on where you're from. While most of America calls a milk shake a milk shake, New Englanders may call it "velvet" or "frappe," or even "cabinet" when in Rhode Island.

The basic ingredients in a milk shake are milk, ice cream, and flavoring, although over the years a variety of ingredients have been used to provide specific flavors, meet certain demands, or minimize costs. For example, some shakes at fast-food restaurants don't even contain milk or ice cream and are formulated to be inexpensive and quick to make.

Although the name is derived from milk, it's the ice cream that makes a milk shake cool and refreshing. To make a shake, milk and flavors are added to ice cream and the mixture is whipped in a blending device. Ice cream already contains air, with up to half of the volume of ice cream made up of small air bubbles. When whipped with milk, even more air is incorporated to make a frothy shake.

Ice cream also contains lots of small ice crystals, which give the cooling sensation and also govern texture. More ice means harder ice cream, and low temperatures mean more ice – that's why ice cream right out of the deep freeze is hard enough to bend a spoon. To make a milk shake that can be sucked through a straw, ice cream needs to be warmed up a bit to melt some ice. Adding milk and whipping are sufficient to turn ice cream into a thick, semi-fluid drink. To make a thicker shake, use more ice cream and less milk so there are more ice crystals.

In a sense, there's a continuum in thickness depending on the amount of ice. Ice cream mix and fluid milk, with no ice crystals, are on the fluid end of the spectrum, whereas deep freeze ice cream, with most of the water in the form of ice, is on the solid end of the spectrum. In between, the milk shake leans toward the more fluid side with less ice, while soft-serve ice cream or custard leans toward the more solid side with more ice. A super thick milk shake is in between the standard milk shake and soft-serve ice cream.

What's your favorite flavor of milk shake? Chocolate, strawberry, and vanilla are traditional favorites, but everything from bubble-gum grape to Cherry Garcia has been tried. There's even a Krispy Kreme flavored milk shake.

One popular milk shake flavoring is malted milk, made by adding malted milk powder to a regular flavored milk shake. Originally developed by the Horlicks brothers in 1873 in Racine, WI as an infant nutritional supplement, malted milk powder is a combination of dried malted barley, wheat flour, and milk. Most sources cite a soda jerk at a Walgreen's soda fountain in Chicago in 1922 as the first person to add malted milk powder to a milk shake to make what's now known as the malted.

Regardless of flavor, ice-cold milk shakes can induce brain freeze, or sphenopalatineganglioneuralgia. The deep cold of the milk shake, or any frozen product, causes the blood vessels in the roof of your mouth to constrict. This is followed by the dilation of the blood vessels to bring heat back into the area when the cold is removed. A neural signal generated by the blood vessel dilation causes a referred pain, meaning a pain felt somewhere other than at the source of the problem, in the head. A brain freeze headache is the result.

The exact location of the referred pain depends on where the cold is applied in the mouth. This might explain why no two brain freeze headaches are exactly the same. Interestingly, researchers also found that brain freeze headaches are only induced in the summer – apparently, drinking a milk shake in the winter doesn't bring on the headache.

To enjoy a milk shake or malted and avoid sphenopalatineganglioneuralgia, it's best to drink slowly and aim the straw away from the roof of your mouth.

54

Circus Peanuts

Circus Peanuts – you either love them or hate them. But why? Do people hate them because of the texture? Or is it because of the flavor? What flavor is it anyway?

The history of Circus Peanuts is clouded, as with most foods, but perhaps for Circus Peanuts it's because nobody wants to admit that they're responsible for developing this much-maligned product. What type of person would come up with the idea of an orange peanut-shaped marshmallow candy with an indeterminate flavor?

These hard orange peanuts, complete with dimpled sides, are considered marshmallow confections. However, Circus Peanuts are much different from the Jet-Puffed marshmallow for roasting on the campfire or bunny-shaped Peeps in the Easter basket. Circus Peanuts, like the marshmallow bits (called marbits) in Lucky Charms, are denser, grained marshmallows, where some of the sugar is found in crystal form.

Probably the main characteristic of marshmallows is the aeration, with a density much less than that of water. That low density means they float in water, or hot chocolate. Even Circus Peanuts float because they're less dense than water, just like the marbits in Lucky Charms that float in your cereal bowl.

Interestingly, the history of the marbits in Lucky Charms started with the Circus Peanut. The story goes that the developer of Lucky Charms, an employee of General Mills, tried shaving some Circus Peanuts into his cereal and loved the effect. A new cereal was born, with the marbits becoming almost legend. Stories abound about the deeper meaning of the marbit shapes, from clover to rainbow. You can even discover your sexual preferences through your likes and dislikes of certain marbits (www.trygve.com/uecharms.html).

Circus Peanuts have a Wisconsin connection, but it's not to the Ringling Brothers circus in Baraboo, WI. The leading manufacturer of Circus Peanuts is Melster Candies, in Cambridge, WI. According to the Food Scientist in charge of quality control and product development at Melster Candies, they make more Circus Peanuts than the other manufacturers combined.

What's in Circus Peanuts and how are they made? The ingredient list in Circus Peanuts includes sugar and corn syrup as the main ingredients, but also listed are gelatin, soy protein, and pectin. There is orange color and an artificial flavor (which one?) added as well.

The sugar and corn syrup provide the sweetness and the gelatin provides the whipping capacity. To make Circus Peanuts, the sugar and corn syrup are mixed and cooked. After the syrup cools a little, the gelatin is added and the syrup is whipped to incorporate air. The aerated syrup, while still warm, is deposited into molds to form the peanut shape. The gelatin sets as the marshmallow cools and holds the air in the candy mass.

The mold for Circus Peanuts is actually a depression in dry corn starch. A tray is filled with corn starch into which depressions are made in the shape of a peanut with the dimples on the side formed in the starch.

Circus Peanuts have two sides. One side is the peanut shape with dimples (the mold side) and the other side (the top) is where the mold is filled. When the marshmallow syrup is poured into the mold, the top side is flat. As the marshmallow cools and dries, some of the sugar in the syrup crystallizes. It's the contraction associated with crystal formation that causes the slight concave depression to form on the top side.

The unique shape comes from the way the candy is formed. The unique flavor comes from...; well, what is the flavor?

It's banana, can you believe it (or even figure it from the taste)? Who would dream up a product that is orange, looks something like a peanut, and tastes like a banana? And what's that got to do with the circus?

55

Marshmallow Peeps

What's your favorite way to eat Peeps? Do you like them fresh out of the box or do you let them dry out a little? Do you find creative ways to play with them?

According to the National Confectioners Association, Peeps are the most popular sugar candy at Easter. Over 70 million of the little marshmallow chicks and bunnies are sold at Easter each year.

The label reads that Peeps are made from sugar, corn syrup, gelatin, colors and flavors, and carnauba wax. Gelatin gives marshmallow the characteristic chewy texture and supports the air bubbles that make it light and fluffy. The sugar and corn syrup form a liquid solution that surrounds the air bubbles. Colors and flavors, of course, provide a unique eating experience.

But what is carnauba wax used for in Peeps?

To make Peeps, gelatin is added to a warm solution of sugar and corn syrup, with air whipped into the mixture as it passes through a beating tube. The aerated marshmallow candy is pressed or extruded out of a nozzle and formed on a conveyor.

To make the little chicks, fluid marshmallow exits a movable nozzle that is moved forward, backward, and up to make the form of a chick. Other Peep forms are simply extruded in the appropriate shape and cut onto a bed of colored sugar crystals.

The final step is the application of the eyes, something that used to be done by hand, but now it is also fully automated. What do you think the eyes are made of?

First, a little Peeps history. Peeps were first made in the 1920s, but were not popularized until 1953 when the Just Born Company began manufacturing them. With improved technology, the time to make a single Peep reportedly has dropped from 27 hours in 1953 to about 6 minutes now.

Are you one of those people who open the package to let the Peeps dry out? Peeps lose moisture when exposed to air with humidity less than about 55 percent, just like our hands on a cold winter day.

Dried out Peeps are hard because the sugary syrup that holds the air bubbles becomes more solid. In a fresh Peep, the sugar syrup is fluid and the marshmallow has a soft, spongy texture. In a dried Peep, the sugar matrix solidifies, sometimes to the point of being like a hard candy. Really, really dried Peeps might then be called petrified Peeps.

Peeps also have interesting properties when heated. Some scientists have 'studied' Peeps when heated, whether in an oven or a microwave, and found that they go through several stages (www.peepresearch.org). When heated at constant pressure, Peeps first expand like a balloon being blown up.

According to the ideal gas law, air in each tiny air bubble must expand with increasing temperature, resulting in an increase in volume. According to these Peep researchers, heating for about 75 seconds in the microwave caused a Peep to nearly double in size.

When heated further, however, the gelatin gel melts, causing the candy mass to flow and leaving a gooey pile of sugary candy where there used to be a Peep. The only recognizable parts of the Peeps that remain after complete melting are the eyes — two forlorn spots in a mass of sugar goo.

As you probably guessed already, the eyes are made from carnauba wax. Carnauba wax is a natural product derived from the leaves of the Carnauba palm tree. It may be used to wax surfboards or M&Ms, as a component of lipstick, or as the eyes of a Peep. The eyes are painted on each Peep right after it has been dusted with sugar.

Now that you know some of the science behind Peeps, perhaps you'll be a more skilled Peeps jouster. Never jousted with Peeps? Arm two Peeps with toothpicks, face them at each other in the microwave and turn on the heat. The ideal gas law kicks in and as they expand, the toothpick from one pricks the other, causing it to burst. Now there's a reason to play with your food.

56

Salt Water Taffy

The heat of summer often means a trip to the beach. Sun, sand, and salt water — taffy, that is. The story goes that an Atlantic City candy maker's shop got flooded in a storm, back in 1883, and the sea water got into all of his candies. When a little girl came in the next day to buy taffy, the shop owner jokingly told her to take some of his "salt water" taffy.

The candy must not have been too bad. Neither was the name, salt water taffy. Now, salt water taffy can be found at beaches up and down the coast, from Florida to Maine. Salt water taffy can even be found in Salt Lake City, where several companies produce the sweet treat.

Does salt water taffy really contain salt water? The salt water taffy we bought recently while on vacation at Cape Cod contained corn syrup, sugar, water, vegetable fat, salt, egg whites, flavors, and colors. Although salt water isn't specifically listed, there is salt and there is water. I suppose that qualifies as salt water.

Salt in candy provides an interesting contrast to the sweetness, but mostly it enhances the taffy flavors. You don't really get a salty taste, but it makes common flavors like chocolate and peanut butter, and even blueberry, taste better.

Watching the candy maker pull taffy is another interesting part of the beach experience. Many small candy makers still pull taffy on a hook mounted on the wall. A good candy maker stretches the candy mass until it's almost falling from the hook and knows exactly when to quickly loop the stretched candy back up over the hook. Besides being good exercise, the continual folding and refolding of the candy mass impart characteristic attributes to the taffy.

First and foremost, pulling taffy incorporates air, which gives the candy a light texture. The candy mass before pulling has a specific gravity of about 1.45 — the dissolved sugars make it denser than water. After pulling, specific gravity may be as low as 1.1, making it easier to eat, less sticky, and more resistant to flow. Does it sink or float in water?

Pulling also modifies consistency, color, and even flavor. Taffy that's not pulled as much has a denser, chewier texture, more like that of AirHeads and old-time Turkish taffy. Some of you may remember Bonomo's ... *Oh, Oh, Oh — its Bonomo's — caaaaandy.* Slam it on the table, it's that hard, and break it into bite-sized, easy-to-eat pieces. French Chew is another example of a hard taffy.

Fresh salt water taffy has a delightful, soft texture that is easy to chew. There's nothing better than biting into a fresh piece of soft taffy. But get into a bag of old taffy and, as one old-time candy maker says, it will steal your teeth — it's so hard to chew.

Most of the toughening during storage is simply due to moisture loss. Except on the most humid of summer days, the water in the taffy wants to escape into the air. The wax paper wrapper on most taffy isn't a very good moisture barrier, as water easily escapes through the twist on both ends. The result of this moisture loss is harder taffy.

What would happen if you warmed up old taffy in the microwave? As long as you only nuked it enough to warm it up, not so hot that it all melted, the taffy would soften to the point where it had nearly the same texture as fresh candy.

So, the two main factors that affect taffy hardness are moisture and temperature. A warm candy with high moisture content is the softest. But, if the taffy is too warm or too high in water content, it wouldn't hold its shape. That's also not good; the little blobs of taffy that stick to the wrapper are difficult to eat.

Getting just the right combination of ingredients, and the proper pulled consistency, is important to making a good salt water taffy — one that won't steal your teeth.

57

Caramel

Do you say kar-ah-mel, kare-ah-mel, kar-mel or kar-mul? According to most dictionaries, either kar-ah-mel or kar-mel is correct, regardless of what aspect of caramel you're trying to describe.

Besides being a fancy word for a shade of brown, caramel can either be a chewy sweet or a burnt sugar concoction used as a flavor or colorant. Caramel can vary in color from a deep brown to a light tan.

How does caramel candy get its color? From caramelization, right?

Wrong.

Do this experiment at home. Put a pair of pots side-by-side on the stove. In one, put a mixture of sugar and corn syrup and in the other, the same mixture of sugar and corn syrup with a little evaporated or powdered milk. Set the burners to medium, stir constantly, and observe the colors that develop as the mixtures cook.

The mixture with milk starts to get brown when the temperature reaches the 225−230°F range, and continues to deepen in color as temperature reaches 240−250 degrees. This is how caramel candies are made.

On the other hand, the sugar−corn syrup mixture doesn't start to turn really brown until the temperature reaches 270−280 degrees. As the temperature exceeds 300 degrees, the color gradually becomes darker and darker. At higher temperatures, even darker colors are formed. This is how caramel colorants and flavorants are produced.

The reaction that gives color development in the simple sugar mixture is called caramelization. The heat causes the sugars to undergo a complex series of reactions, with the end result being the formation of volatile flavors and polymeric caramellans. This type of caramel is added to colas and soy sauce, among other foods, to provide color and flavor.

The caramel candy mixture, sugar, and corn syrup with evaporated milk turns brown at lower temperatures because certain sugars react with the milk proteins. This is the same reaction — Maillard browning — that turns toast brown and gives raisins their brown color. Interestingly, despite the name, caramel candy doesn't get its color from the caramelization reaction.

How is caramel on a stick different from the gooey caramel inside a chocolate-covered candy? It's mostly due to the water content, governed by the temperature to which the caramel mass is cooked. Cooking to higher temperatures, like 270 or 280 degrees, leads to a dark brown caramel and also reduces the water content. At lower water content, the caramel, which is an amorphous sugar–protein matrix with small fat globules dispersed throughout, is extremely viscous, has a firm texture, and stands up to its own weight when cooled.

Caramels cooked to lower temperatures not only have a lighter color, but they also contain more water. This gives them a soft runny characteristic. Since the higher water content gives a less viscous mass, they exhibit cold flow — the ability of an amorphous matrix to gradually flow at room temperature, eventually forming a puddle of caramel.

Caramels can also have what's called a "short" texture. To demonstrate this, find a standard commercial caramel, the ones that come in the shape of a cube, and a fresh homemade caramel. Grab the homemade caramel by two ends and slowly pull it apart. It stretches and stretches as you separate your hands, until eventually the long caramel string breaks. This is characteristic of a chewy caramel.

Now do the same with the commercial caramel. It should only stretch by an inch or two before the strand breaks. This is the "short" characteristic, caused by the presence of small sugar crystals that break up the stretchy protein–sugar strands.

So, regardless of whether you say kar-ah-mel or kar-mel, rest assured that candy makers go to great lengths to make their candy just the way you like it: dark or light, hard or soft, short or chewy.

Just don't ask for a caramel in England — they call it toffee there.

58

Life Is Like a Box of Chocolates

Forrest Gump's momma says, "Life is like a box of chocolates, you never know what you're gonna get." As the holidays approach, this statement is more pertinent than ever. How do you know what's inside each one of those little chocolate delights?

Chocolate manufacturers know what's inside each of their pieces, but most don't provide a scorecard to help you figure it out. With a little information, you can tell without resorting to a fingernail in the bottom of each candy.

Look carefully at each piece. Chocolate makers distinguish their different candies in two ways, by shape and by the swirly design on top. Chocolate-covered cherries are easy to spot; they have a distinct rounded shape and may be foil wrapped — that's just in case there are "leakers." Creams are similar in shape to the cherries, although usually a little smaller, but you can't distinguish the flavor of the cream by the shape. You're as likely to get maple flavor as raspberry cream.

Caramels, nuts, nougats, and others tend to be either square or rectangular in shape and are even more difficult to distinguish. That's why the chocolate makers swirl the chocolate on top in a distinctive pattern, like a curlicue, which is different for each piece. Each cream flavor has a slightly different swirl that indicates its unique flavor.

How are these chocolate pieces made? How do they get those gooey, soft centers inside the layer of chocolate? Here's where the science and technology comes in. There are several ways to do it.

A center that's firm enough to stand on its own, like a firm caramel, can be passed through a curtain of chocolate in an enrober.

The solid caramel first passes across a roller coated with chocolate to coat the bottom. The partially coated piece is then sent through a yummy curtain of chocolate to enrobe the rest of the piece. The coated piece passes through a cooling tunnel to solidify the chocolate and it's ready for packaging.

For centers that are too soft to stand on their own, like a gooey caramel, a molding process can be used. To make a chocolate shell, melted chocolate is poured into a mold (with the proper swirl pattern at the bottom); it's turned upside down to shake out any excess chocolate and allowed to cool. Once this chocolate shell has solidified, a fluid center can be pumped into the mold. To close off the bottom, more melted chocolate is poured on top of the mold and any excess is scraped off. After cooling, the pieces are removed from the mold and packaged.

A new technology — one-shot depositing — has made it even easier to make chocolates filled with soft centers. In one hopper is melted chocolate and in the other is a fluid candy center, whether caramel or cream. The nozzle for the center filling is nestled within the nozzle for the chocolate, like nested measuring cups. The depositor is sequenced so that the chocolate flow starts a fraction of a second before the center flow, so the mold gets coated with chocolate first. The candy center then flows within the chocolate to give the filling. The filling flow stops a fraction of a second sooner than the chocolate to give a completely chocolate-coated candy piece.

Cordial cherries, a gooey cherry center inside a shell of chocolate, use a secret ingredient — invertase. It's an enzyme that breaks down sucrose into fructose and glucose. When first made, the center of a cordial cherry is a hard fondant (harder even than a Peppermint Pattie). The solid fondant piece, with a cherry inside, is firm enough to get enrobed in chocolate, after which it is cooled and packaged.

Over the following two weeks, the invertase breaks down the sucrose, resulting in a soft, gooey center completely coated with chocolate. If you happen to get a hard Cordial cherry center, it either hasn't been aged long enough or that particular piece didn't get enough invertase.

So why is life like a box of chocolates? Is it because you don't know what a person is like until you get to know them? Or, is it because you never know what to expect from each new experience in life?

Or is that bumper sticker right? Life is like a box of chocolates — full of nuts!

59

Hollow Chocolate Bunnies

One of the mainstays in the Easter basket is the chocolate bunny. Some are hollow, some solid. Some are small, some huge.

Are you one of those people who eat the ears off first? Seems like most people do, perhaps because they're the easiest part to bite into. But, biting the ears off a large hollow chocolate bunny often makes a mess, with chocolate shattering into small pieces that leave stains on your clothes. Eating large solid bunnies, though, isn't much easier since you have to sort of gnaw on their edges to bite off some chocolate.

Large solid bunnies may be a little harder to eat, but they're easier to make. The candy maker snaps together a two-sided mold, each side half a bunny, and then fills the mold with tempered chocolate through a hole at the bottom. After cooling, the mold is popped open to release the bunny within.

Making hollow chocolate bunnies is a little trickier. Again, the candy maker fills the mold with chocolate, but, in this case, the mold is turned over after a few minutes of cooling. The remaining liquid chocolate is shaken out onto a tray, leaving a wall of solidified chocolate at the mold surface. The mold is then placed firmly on the pool of chocolate just shaken from the mold to form the bottom of the bunny. A good chocolate maker can produce perhaps 50–75 bunnies per hour.

To increase production, at least for the smaller bunnies, chocolate makers use a mold spinner, a machine that can produce up to 720 bunnies per hour depending on the size of the bunny. The bunny molds are magnetically clamped to the outside of a rotating drum. As the drum rotates, each mold spins 360 degrees around on its pivot, setting up a multi-circular motion that ensures the chocolate coats the entire inner surface of the mold without pooling in any particular area.

This complicated motion yields a uniform coating of solidified chocolate on the inside of the mold. After the chocolate cools and sets, the mold is removed from the machine and popped open to release the hollow bunny. The thickness of the bunny depends on how much chocolate was poured into the mold.

To color the ears, eyes or other body parts, the bunny maker either applies chocolate paint to the inside of the molds before putting the two halves together or paints them on after the bunny is made. White chocolate, essentially milk chocolate without the cocoa, is colored with food coloring to make white eyes, pink ears or colored clothes. When the tempered chocolate is poured into the mold, it melts a little of the colored chocolate prepainted on the mold, so that the color sticks to the chocolate bunny when the mold is opened.

Hollow chocolate bunnies are fragile, especially the big ones. They're prone to breakage when dropped or bounced during shipping. What's the best way to package hollow bunnies to protect them from accidental falls?

Researchers in the University of Missouri-Rolla packaging program actually studied the problem of protecting hollow chocolate bunnies from accidental falls. They concluded that the bunny needed a cushioned ride to survive a fall. A highly sophisticated, air suspension package was recommended to prevent breakage – sort of like air bags for the vulnerable bunny. Unfortunately, despite these recommendations, chocolate bunny safety is callously disregarded since most hollow bunnies are still sold either in plastic wrap or metal foil. They're still susceptible to accidental damage.

What's the largest hollow chocolate bunny made? One New York company makes one, by hand, that stands over two feet tall, weighs about ten pounds, and costs $80. The company doesn't ship it by mail because it's too fragile. Even an air-cushioned ride wouldn't protect this mammoth rabbit.

However, rumor has it that there's a three-foot tall hollow bunny weighing 20–25 pounds out there somewhere. Hollow chocolate bunny hunters have been searching for this lunker for years. The last sighting was at a nut and chocolate company in Michigan, although its existence cannot be confirmed.

Imagine trying to eat the ears off that monster.

60

Chocolate Gone Bad

What do chocolate, the Dogon statues from Mali, cosmetic pencils, and flowers have in common? They can all exhibit a form of bloom, which, except in flowers, is a surface blemish composed of small spots or a white haze that forms under certain conditions.

Bloom appears as a white haze on the surface of the chocolate. Actually, there are several different kinds of bloom. Sugar bloom occurs when chocolate gets a little moist. For example, when you remove a piece of chocolate from the refrigerator or freezer on a humid day, moisture in the air can condense on the surface of the chocolate. The water dissolves some of the sugar and later, when the water dries off, the sugar remains as a white spot – sugar bloom.

Probably more common is fat bloom, of which there are several types. One readily apparent form of bloom occurs when chocolate is stored at warm (and fluctuating) temperatures for too long. Over time, a white haze builds up on the surface. Fat bloom looks a little like mold, but it's only cocoa butter crystals growing out of the surface of the chocolate.

Many years ago, we had a chocolate Santa with only the head eaten off. We put the body of Santa back in the box and placed it on top of the refrigerator. When we found it again in June, the entire Santa had a white layer that had initially formed on the surface, but by then had extended partway into the body. Santa's interior regions were still nice chocolate, but exterior, it looked like snow had gotten into and under Santa's skin.

To understand fat bloom, first we have to understand the structure of chocolate. A normal chocolate bar contains about 50 percent sugar in the form of very small crystals – so small we can't feel them in our mouths. Sugar adds sweetness. If you've ever eaten unsweetened chocolate, you might wonder how it got the reputation as the food of the gods. Then there are about 16 percent cocoa solids, the remains of ground up cocoa beans. The rest, about 32–35 percent, is cocoa butter, the fat from the cocoa bean.

Cocoa butter is a natural fat with melting properties that make it ideal for chocolate. At room temperature, it's mostly solid, imparting a desirable snap to a chocolate bar. In your mouth, cocoa butter melts completely, giving chocolate its creamy, smooth texture when we eat it.

In well-made chocolate, cocoa butter crystals are very small. The fat crystals at the surface reflect light, giving a shiny, glossy appearance. It's when chocolate isn't solidified properly or goes through abusive storage that the cocoa butter crystals get larger. When temperatures fluctuated on top of our refrigerator, the cocoa butter crystals rearranged within the chocolate Santa, and large crystals grew out of the surface to give the white hazy appearance.

Bloomed chocolate, with the white surface, isn't bad for you; it just isn't as appealing nor does it taste quite as good. But it's not mold, it's just cocoa butter crystals.

Fat bloom also occurs in materials other than chocolate – a similar phenomenon disfigures the surfaces of some cosmetics and the Dogon statues from Mali. Cosmetic pencils have been known to grow bloom, although the problem is quite rare so not much of a concern.

On the other hand, museum curators are extremely concerned about surface disfiguration on some historic relics. The Dogon statues, from Mali in Africa, are wooden carvings that were infused with fats. Over time, fat crystals have grown out of the surface of these statues, in much the same way as our chocolate Santa, causing an undesirable appearance.

Neither museum curators nor chocolate makers know how to prevent fat bloom. It remains a well-studied, but still unsolved, mystery.

Printed in the United States of America